SpringerBriefs in Computer Science

SpringerBriefs present concise summaries of cutting-edge research and practical applications across a wide spectrum of fields. Featuring compact volumes of 50 to 125 pages, the series covers a range of content from professional to academic.

Typical topics might include:

- A timely report of state-of-the art analytical techniques
- A bridge between new research results, as published in journal articles, and a contextual literature review
- A snapshot of a hot or emerging topic
- An in-depth case study or clinical example
- A presentation of core concepts that students must understand in order to make independent contributions

Briefs allow authors to present their ideas and readers to absorb them with minimal time investment. Briefs will be published as part of Springer's eBook collection, with millions of users worldwide. In addition, Briefs will be available for individual print and electronic purchase. Briefs are characterized by fast, global electronic dissemination, standard publishing contracts, easy-to-use manuscript preparation and formatting guidelines, and expedited production schedules. We aim for publication 8–12 weeks after acceptance. Both solicited and unsolicited manuscripts are considered for publication in this series.

**Indexing: This series is indexed in Scopus, Ei-Compendex, and zbMATH **

Renzhi Yuan · Zhifeng Wang

Non-Line-of-Sight Ultraviolet Communications

Principles and Technologies

 Springer

Renzhi Yuan
School of Information and Communication
Engineering
Beijing University of Posts
and Telecommunications
Beijing, China

Zhifeng Wang
Department of Electrical and Electronic
Engineering
The University of Hong Kong
Hong Kong, China

ISSN 2191-5768   ISSN 2191-5776 (electronic)
SpringerBriefs in Computer Science
ISBN 978-981-97-8542-1   ISBN 978-981-97-8543-8 (eBook)
https://doi.org/10.1007/978-981-97-8543-8

This Springer imprint is published by the registered company Springer Nature Singapore Pte Ltd.
The registered company address is: 152 Beach Road, #21-01/04 Gateway East, Singapore 189721,
Singapore

If disposing of this product, please recycle the paper.

Preface

Thought enjoying high transmission rate, large capacity, and good security compared with radio-frequency communications, traditional wireless optical communications in infrared or visible wavebands are mainly restricted in line-of-sight (LOS) communication scenarios. Ultraviolet (UV) communication working in "solar-blind" wavebands can achieve non-line-of-sight (NLOS) communication links thanks to the strong scattering effects of UV signals traveling in the atmosphere. Besides, the UV communication also enjoys inherent advantages such as low background noise, high local security, flexible networking, and good weather adaptivity, making it a promising both civil and military communication technology in electromagnetic-sensitive environments, such as high-precision instruments factory and electromagnetic suppressed battlefields.

In this book, we introduce the basic idea and key technologies of NLOS UV communications, including the channel modeling, achievable information rate, full-duplex UV communications, relay-assisted UV communications, and NLOS UV positioning. We also introduce some promising research directions of UV communications such as the integrated UV communication and positioning, UV spatial diversity techniques, and NLOS UV networking. The potential readers of this book can be graduate students or young researchers who are interested in UV communications or working in wireless optical communication areas. We hope this book can provide a comprehensive understanding of NLOS UV communications and stimulate some new ideas in the development of practical UV communications.

At last, we would like to thank the support from the State Key Laboratory of Networking and Switching Technology in Beijing University of Posts and Telecommunications and the support from the National Natural Science Foundation of China (No. 62201075).

Beijing, China   Renzhi Yuan
August 2024   Zhifeng Wang

Contents

Chapter 1
Introduction to Ultraviolet Communications

Abstract Optical wireless communications (OWCs) employing electromagnetic waves in optical wavebands as information carriers can achieve higher communication bandwidth compared with radio frequency based wireless communication. However, the good directionality of optical waves degrades its non-line-of-sight (NLOS) transmission ability for avoiding obstacles. Ultraviolet (UV) communications working in "solar blind" wavebands (200–280 nm) can well overcome this drawback of OWCs and achieve NLOS links due to the strong scattering effect of the ultraviolet lights passing through the atmosphere. In this chapter, we first present the basic idea and potential applications of UV communication in Sect. 1.1. Then we introduce the research history of UV communication in Sect. 1.2. Then we introduce the key technologies evolved in UV communications in Sect. 1.3. In Sect. 1.4, we summarize this chapter and introduce the structure of this book.

Keywords Ultraviolet communication · Non-line-of-sight · Scattering effect

1.1 Basic Idea and Potential Applications of Ultraviolet Communications

1.1.1 Basic Idea of Ultraviolet Communications

The ultraviolet (UV) communication employs the UV light in "solar blind" waveband (200–280 nm) as information carriers [1]. The name "solar blind" is derived from the fact that the solar radiation in 200–280 nm is strongly absorbed by the Ozone layer of the atmosphere such that little UV lights can reach the earth's surface. Therefore, UV communications enjoy low background radiation noise compared with other optical wireless communications [2]. Besides, the strong absorption effect of UV lights in the atmosphere also guarantees a good local security due to the verified low-probability-of-detection [3]. Therefore, the UV communication becomes a promising non-line-of-sight (NLOS) optical wireless communication technology and attracted increasing research attentions in recent decades.

R. Yuan and Z. Wang, *Non-Line-of-Sight Ultraviolet Communications*,
SpringerBriefs in Computer Science, https://doi.org/10.1007/978-981-97-8543-8_1

However, the risk of potential UV radiation exposures on human bodies restricts the applications of UV communications and early studies of UV Communications were focused mainly on its military applications [2]. Fortunately, the fast development of the industrial internet spurs new opportunities to UVC because human operators in future industrial internet may be replaced by robots. Besides, similar to other optical wireless communications (OWCs), such as the visible light communication and the infrared communication, the UV communication can also work under line-of-sight (LOS) links. The LOS UV communication can achieve a higher capacity compared with other OWCs due to its higher bandwidth and lower background noises; and therefore, it can be used in some special communication scenarios, e.g., the ultra-fast data exchange among super computers. Since the NLOS ability of UV communication is unique compared with other OWCs, we will focus on the NLOS UV communication in this book.

A typical NLOS UV communication system is shown in Fig. 1.1 [1]. The transmitter emits a UV light beam and the UV lights are scattered by molecules and aerosols in the atmosphere. Then the scattered UV lights in the receiving field-of-view (FOV) are detected by the receiver. The basic idea and the experimental demonstration of UV communications were proposed in 1960s [2]. However, due to the lack of compact UV light sources and UV detectors, the UV communication did not become a promising technology until the 21st century. In 2002 and 2007, the US defense advanced research projects agency (DARPA) launched the semiconductor UV optical source program and the deep ultraviolet avalanche photodiode program, which greatly promoted the development of UV light-emitting diode (LED) sources and UV detectors [1]. Since then, the development of a compact UV transceiver based on UV LEDs and UV detectors becomes possible and the research of UVCs attracted increasing attentions [1].

Fig. 1.1 A typical "Solar-blind" NLOS UV communication system

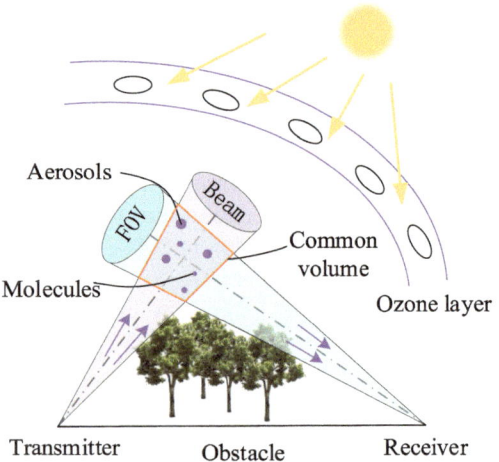

Table 1.1 Comparison between UVC, VLC, IRC, and RF-WC

Types	OWC			RF-WC
	UVC	VLC	IRC	
Wavebands	200–280 nm	350–780 nm	750 nm–1 mm	>1 mm
Range	+	++	+++	++++
Capacity	++++	+++	++	+
NLOS ability	+++	−	−	++++
Local security	++++	++	+	−
Anti-electromagnetic interference	++++	+++	++	+
Climate adaptability	+++	+	+	++++
Human safety	+	+++	++++	++++

A qualitative comparison between the UV communication (UVC), visible light communication (VLC), infrared communication (IRC), and radio frequency based wireless communication (RF-WC) in terms of communication range, achievable capacity, NLOS ability, local security, anti-electromagnetic interference, climate adaptability, and human safety are summarized in Table 1.1. From Table 1.1 we can observe that the major advantages of OWCs compared with the RF-WC are high capacity and good anti-electromagnetic interference ability. Besides, we can see that the major advantages of the UV communication compared with other OWCs are the NLOS ability and the good climate adaptability, where the latter is due to the enhanced scattering effects of UV lights under weather conditions with low-visibility [1]. The major drawbacks of the UV communication are the short communication range and the risk of human safety. Therefore, the UV communication is suitable for applications with high requirements for capacity, NLOS ability, local security, anti-electromagnetic interference, and good climate adaptability, while with low requirements for communication range and human safety.

1.1.2 Potential Applications of Ultraviolet Communications

The strong attenuation of UV lights traveling in the atmosphere yields a good local security of the UV communication since any detecting attempt of the adversary cannot be achieved from a distance. Besides, the UV communication can work under complex communication environments due to its NLOS ability and it is immune to the electromagnetic interference. These inherent advantages made UV communication a promising option for military secure communications, e.g., the secure communication among aircraft carrier fleet in the ocean, combat units on the ground, or UAVs in the air. Moreover, the UV communication can also be used in the high-speed communications under electromagnetic sensitive environments, e.g., the high-precision

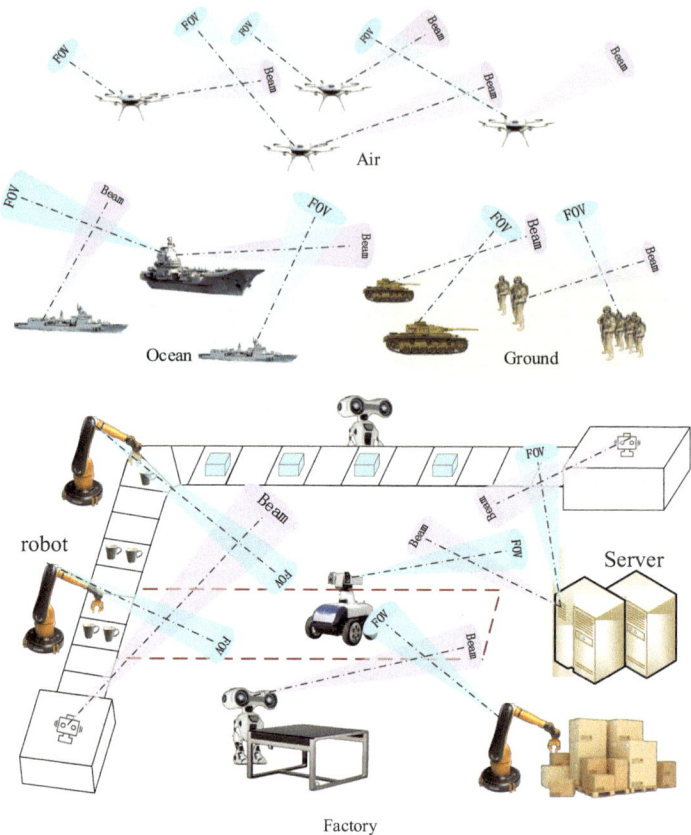

Fig. 1.2 Potential applications of UV communications

equipment factories and the medical equipment factories. The major applications of UV communication are summarized in Fig. 1.2.

1.2 Research History of Ultraviolet Communications

The pioneering research of UV communication can date back to 1960s. In 1964, G. L. Harvey put forward some key technical issues of UV communication [1]. In 1968, D. E. Sunstein built the first experimental platform for UV communications. In 1985, M. Gelleret established a relatively mature ultraviolet communication system by utilizing a mercury lamp as the light source and a photomultiplier tube combined with a filter as the receiver, and verified the feasibility of UV communication for the first time [1]. However, early UV communication systems employed gas discharge lamp as the light source, which suffers from long response time and alrge volume.

Fig. 1.3 Typical
performance of some
reported UV communication
systems

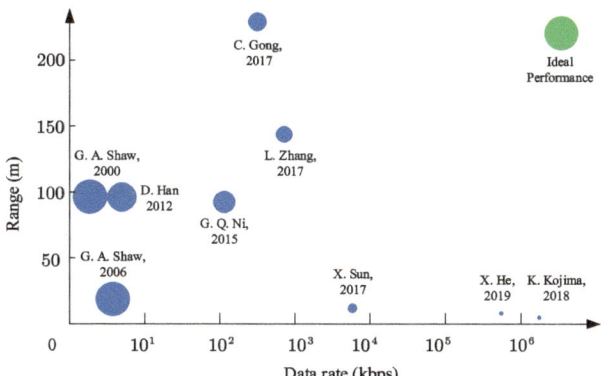

In 2002, the US defense advanced research projects agency (DARPA) launched the
semiconductor ultraviolet optical Source (SUVOS) program; and in 2007, DARPA
launched the deep ultraviolet avalanche photodiode (DUVAP) program. After then,
the commercial UV LED and UV detectors with compact size become available,
which promote the study of UV communications after 2010s.

In recent years, considerable works have been performed on the NLOS UV channel modeling [4–9], channel estimation [10–13], coding and modulation [14–16],
diversity reception [17–19], duplex and relay techniques [20–23], and experimental
tests [24–29]. The major performance metrics for UV communications include the
communication range, the data rate, and the NLOS ability. Figure 1.3 summarized
the performance of some typical UV communication systems. The NLOS ability is
denoted by the radius of the data point. From Fig. 1.3 we can see that early experimental results for UV communications can achieve NLOS communication within
100 m but with Kbit/s data rate. The fast development of UV LED and UV detectors promoted the experimental UV communication performance in recent years.
UV communication systems with Mbit/s or even Gbit/s were reported under short
communication ranges. Generally, an ideal UV system enjoys large communication
range, high data rate, and good NLOS ability is still absent.

1.3 Key Technologies of Ultraviolet Communications

The key technologies of UV communications include the channel modeling, channel
estimation, coding and modulation, diversity reception, capacity analysis, duplex
communication, relay-based communications, etc. In this book, we mainly focused
on the channel modeling, capacity analysis, duplex communication, and relay-based
communication. Besides, we also introduce the NLOS UV positioning techniques,
which is a promising technique for improving the channel performance of UV
communications.

1.3.1 Channel Modeling

To explore the capacity of UV communications, we need first to establish the channel model of NLOS UV communications. How to estimate the channel path loss and the channel impulse response (CIR) of a given UV communication channel link is the key to the channel modeling of UV communications.

The single-scattering model is the first proposed channel model for UV communications, where the photon is scattered only once before arriving the receiver. The analytical single-scattering model can be expressed as a triple integral on the common volume determined by the transmitting beam and the receiving field-of-view. However, to calculate the accurate path loss and CIR of the single-scattering model, it is necessary to find the integral limits on the common volume, which is a challenging task in most transceiver geometries. In recent years, some approximated single-scattering channel models were proposed to simplify the calculation of the single-scattering model. However, these approximated models are proposed under certain geometrical constrains and thus cannot be applied to arbitrary system geometries.

As the communication range or transmitting or receiving elevation angle increases, multiple scattering effects become pronounced. Therefore, it is necessary to establish channel models for multiple scattering channels in long-distance communication ranges. The multiple scattering models are usually based on Monte-Carlo methods, which often suffer from low computational efficiency. Therefore, how to improve the computational efficiency of multiple scattering models for UV communications is the key research direction for the channel modeling of long-distance UV communications.

1.3.2 Achievable Information Rate

The analysis of the channel capacity for NLOS UV communications under power constrains is still an open research issue. However, if the modulation scheme is prefixed, we can derive the achievable information rate (AIR) of UV communication under given modulations. Currently, the research of capacity analysis for UV communications mainly focused on the AIR analysis under either on-off keying (OOK) or pulse-position modulation (PPM) schemes.

Since the NLOS UV communication is based on the scattering effect of UV signals travelling in the atmosphere, there exists severe inter-symbol-interference (ISI) between adjacent bits when the transmission rate is high enough. This ISI effect will inevitably deteriorate the UV communication performance. Therefore, there exist an appropriate transmission rate such that the information bit rate achieves its largest value.

The amount of the ISI is closely related to the shape of the CIR. Different transceiver geometries correspond to different shapes of CIR, and thus result in different ISI. Therefore, the difficulty of analyze the AIR of NLOS UV communications comes from the complexity of the transceiver geometries. Besides, the modulation scheme will also affect the AIR. How to analyze the AIR under various modulation schemes is an important issue in UV communications.

1.3.3   UV Duplex Communications

Typical duplex schemes for optical wireless communications include the time-division duplex (TDD), the wavelength division duplex (WDD), and the space-division duplex (SDD). Compared with the TDD and WDD, the SDD scheme can achieve the full-duplex communications by sharing the same time-frequency resource between the transmitting link and the receiving link. However, the traditional OWCs, such as IRC and VLC, require the alignment between the transmitter and the receiver, which greatly restricts the application scenarios of the full-duplex optical communications. The NLOS ability of UV communication relaxes the restriction of transceiver alignment for full-duplex UV communications. Therefore, the NLOS UV communication can enjoy a much more flexible full-duplex communication links compared with the traditional OWCs.

Due to the good directionality of the infrared and visible lights, traditional OWCs can achieve the full-duplex communication by simply separating the transmitting link and the receiving link in the space. However, due to the strong multiple scattering effects of UV lights in the atmosphere, the NLOS full-duplex UV communication using the SDD technique suffers serious self-interference between the transmitting link and the receiving link in practical implementations, which will greatly degrade the full-duplex performance. Besides, the NLOS UV communication usually undergoes serious path losses and the received signal strength is usually much weaker than that of traditional OWCs, which further aggravates the influences of self-interference. Therefore, the performance of the full-duplex UV communication using the SDD technique may be worse than that of the half-duplex UV communication using the TDD technique due to the serious self-interference.

In radio-frequency (RF) communications, the self-interference can be cancelled if the self-interference is perfectly estimated. However, the processing of self-interference for NLOS UV communications is fundamentally different from RF communications because only the average self-interference can be estimated due to the granularity of the photons; and we cannot completely cancel the self-interference in the NLOS full-duplex UV communications. Therefore, it is important to explore interference cancelling methods for mitigating the influence of self-interference in NLOS full-duplex UV communications.

1.3.4   Relay-Assisted UV Communications

Due to the strong scattering and absorption effects of UV signals in the atmosphere, UV communications suffer from high path losses, which restricts the UV communication range and degrades its performance in practical applications. The relay-assisted communication is a typical technology to mitigate the influence of high path losses and expand the communication range, which can enhance the flexibility of UV communications under complex environments. For example, the relay-assisted UAV can achieve high capacity and robust UV communication links in mountains and forests. Therefore, it is necessary to employ relay technology in ad-hoc UV communication networks. For example, it was reported that a half-duplex relay can achieve longer communication distances and better communication reliability compared with the direct communication system.

However, the half-duplex relay-assisted UV communication will result in high transmission delay and low efficiency of time-frequency utilization, especially for multi-hop relay systems. The full-duplex relay allows receiving and forwarding the signal at the same time slot and provides higher transmission rates compared with the half-duplex relay system. However, due to the strong scattering and reflecting effects of UV signals in the atmosphere, the full-duplex relay will suffer inevitably serious inter-relay-interference (IRI), which will greatly degrade the performance of multi-hop UV communication systems. Therefore, it is necessary to investigate the influence of interferences on the communication performance and the interference mitigating methods for the full-duplex relay assisted UV communications.

Existing works related to the relay assisted UV communications mainly focus on the half-duplex relay schemes. However, the half-duplex relay suffers from high transmission delay and low efficiency of time-frequency utilization, especially for multi-hop UV communications. It is proved that full-duplex relaying communications can achieve higher potential performance gains than half-duplex relaying communications. The full-duplex relay was studied for a multi-hop UV communication system, where the relays simultaneously detect and transmit the UV signals. The IRI due to the single scattering effect between different relays was considered. However, the self-interference (SI) due to the multiple scattering effects and the reflecting effects between the detection link and the forward link for each relay was ignored most existing studies. For the full-duplex relay based multi-hop UV communication systems, both the IRI and the SI need to be considered; and the relay placement and the transmit power can be optimized to mitigate the impacts of the interference.

1.3.5   NLOS UV Positioning

Optical positioning techniques have inherent advantage such as high precision, low power consuming and immunity to the electromagnetic interference. Traditional optical positioning techniques, such as visible light positioning and infrared positioning, require line-of-sight links between the user and the target, which will fail when only NLOS links are available and thus greatly limit the application scenarios of optical positioning techniques. Therefore, it is meaningful to study the optical positioning technique capable of NLOS positioning.

The UV signal can establish NLOS links between the user and the target due to the strong scattering effects of UV light traveling in the atmosphere. Besides, the UV signal is also influenced by the strong absorption effects traveling though the atmosphere, which gives the UV signal two brand-new advantages. The first advantage is the local security depending on the fact that potential eavesdroppers have low possibility to detect the UV signal beyond a long distance. The second advantage is the low background noise because the Ozone layer can strongly absorb the background solar radiation in the UV waveband. These advantages make UV signals suitable for both communication and positioning in NLOS scenarios. Currently, the UV signals are primarily utilized for communication purposes. The study of the UV positioning technique can improve the robustness of NLOS UV communications. Specifically, when the receiver has some prior information of the transmitter, such as the location and the pointing directions, the receiver can adjust its receiving geometry to achieve a better link quality and thus improve the performance of NLOS UV communications.

Some preliminary works related to UV positioning were performed in the context of UV mesh networks. These works simply employ traditional wireless triangulation positioning method, which require the locations and receiving powers of multiple receivers. However, in practical UV communication systems, the user usually knows its own location and receiving power. The UV positioning using one single user was first studied in neighbor discovery protocols of UV ad-hoc networks. However, these neighbor discovery protocols has some restrictions on the pointing direction of the receiving FOV, e.g., the receiving FOV of all users are required to be placed vertically, which greatly restricts the application scenario of the neighbor discovery protocols. In practical UV communication systems, the transmitter can point at arbitrary directions. Therefore, it is necessary to study the NLOS UV positioning method that suitable for arbitrary pointed transmitters.

1.4 Summarization and Organization of This Book

The NLOS UV communication enjoys inherent advantages including low background noise, high local security, good weather adaptability, and NLOS communication ability, which makes it a promising communication technology for future secure networking among unmanned vehicles. In this book, we introduce the basic idea, key technologies, and future directions of NLOS UV communications. We mainly focus on the key technologies including channel modeling, achievable information rate analysis, full-duplex UV communication, relay-assisted UV communication, and NLOS UV positioning. The rest of this book is organized as follows:

- In Chap. 2, we introduce the channel modeling of NLOS UV communications, including analytical channel models and stochastic channel models. The channel path loss and the channel impulse response are derived and numerically analyzed.
- In Chap. 3, we introduce the achievable information rate (AIR) of NLOS UV communications under inter-symbol-interference channels. The AIRs for OOK and PPM modulations are derived and numerically analyzed.
- In Chap. 4, we introduce the full-duplex UV communication technology. The modeling and performance analysis for full-duplex UV communication is presented. The communication performance of full-duplex UV communication and simplex UV communication is compared.
- In Chap. 5, we introduce the relay-assisted UV communication to increase the communication range of UV communications. A joint optimization of both relay placement and transmit power is introduced to mitigate the influence of inter-relay-interference.
- In Chap. 6, we introduce the NLOS UV positioning technology to help the transmitter maintain stable single-scattering links. An NLOS UV positioning method based on receiving power is presented and is extended to the case with full-duplex communications.
- In Chap. 7, we introduce some future prospects of NLOS UV communications on research directions including directions of integrated UV communication and positioning, spatial diversity, and UV networking.

References

1. Yuan, R., Ma, J.: Review of ultraviolet non-line-of-sight communication. China Commun. **13**(6), 63–75 (2016)
2. Vavoulas, A., Sandalidis, H.G., Chatzidiamantis, N.D., et al.: A survey on ultraviolet C-band (UV-C) communications. IEEE Commun. Surveys Tuts. **21**(3), 2111–2133 (2019)
3. Weisman, M.J., Dagefu, F.T., Moore, T.J., et al.: Analysis of the low-probability-of-detection characteristics of ultraviolet communications. Opt. Express **28**(16), 23640–23651 (2020)
4. Luettgen, M.R., Reilly, D.M., Shapiro, J.H.: Non-line-of-sight single-scatter propagation model. J. Opt. Soc. A. A **8**(12), 1964–1972 (1991)
5. Wang, L., Xu, Z.Y., Sadler, B.M.: An approximate closed-form link loss model for non-line-of-sight ultraviolet communication in noncoplanar geometry. Opt. Lett. **36**(7), 1224–1226 (2011)
6. Wu, T.F., Ma, J.S., Yuan, R.Z., et al.: Single-scatter model for short-range ultraviolet communication in a narrow beam case. IEEE Photon. Technol. Lett. **31**(3), 265–268 (2019)
7. Yuan, R.Z., Ma, J.S., Su, P., et al.: An integral model of two-order and three-order scattering for non-line-of-sight ultraviolet communication in a narrow beam case. IEEE Commun. Lett. **20**(12), 2366–9 (2016)
8. Drost, R.J., Moore, T.J., Sadler, B.M.: UV communications channel modeling incorporating multiple scattering interactions. J. Opt. Soc. A. A **28**(4), 686–695 (2011)
9. Yuan, R.Z., Ma, J.S., Su, P., et al.: Monte-Carlo integration models for multiple scattering based optical wireless communication. IEEE Tran. Commun. **68**(1), 334–348 (2020)
10. El-Shimy, M.A., Hranilovic, S.: Binary-input non-line-of-sight solar-blind UV channels: modeling, capacity and coding. J. Opt. Commun. **4**(12), 1008–1017 (2012)

11. Gong, C., Xu, Z.: Channel estimation and signal detection for optical wireless scattering communication with inter-symbol interference. IEEE Trans. Wirel. Commun. **14**(10), 5326–5337 (2015)
12. Wei, Z., Hu, W., Han, D., et al.: Simultaneous channel estimation and signal detection in wireless ultraviolet communications combating inter-symbol-interference. Opt. Express **26**(3), 3260–3270 (2018)
13. Hu, W., Wei, Z., Popov, S., et al.: Noncoherent detection for ultraviolet communications with inter-symbol interference. J. Light. Techol. **38**(17), 4699–4707 (2020)
14. Noshad, M., Brandt-Pearce, M., Wilson, S.G.: NLOS UV communications using M-ary spectral-amplitude-coding. IEEE Trans. Commun. **61**(4), 1544–1553 (2013)
15. Cao, T., Wu, T., Pan, C., et al.: A power-domain MST scheme with BPPM in MLOS ultraviolet communications. IEEE Photon. J. **15**(1), 1–10 (2023)
16. Cao, T., Wu, T., Pan, C., et al.: Performance of multipulse pulse-position modulation in MLOS ultraviolet communications. IEEE Commun. Lett. **27**(3), 901–905 (2023)
17. Qin, H., Zuo, Y., Li, F., et al.: Noncoplanar geometry for mobile NLOS MIMO ultraviolet communication with linear complexity signal detection. IEEE Photon. J. **9**(5), 1–12 (2017)
18. Yuan, R., Peng, M.: Single-input multiple-output scattering based optical communications using statical combining in turbulent channels. IEEE Trans. Wirel. Commun. **23**(4), 2560–2574 (2023)
19. Wang, S., Peng, M., Yuan, R.: MIMO free-space optical communications using photon-counting receivers under weak links. IEEE Commun. Lett. **27**(4), 1185–1189 (2023)
20. Ardakani, M.H., Heidarpour, A.R., Uysal, M.: Performance analysis of relay-assisted NLOS ultraviolet communications over turbulence channels. J. Opt. Commun. Netw. **9**(1), 109–118 (2017)
21. Gong, C., Wang, K., Xu, Z., et al.: On full-duplex relaying for optical wireless scattering communication with on-off keying modulation. IEEE Trans. Wirel. Commun. **17**(4), 2525–2538 (2018)
22. Refaai, A., Abaza, M., El-Mahallawy, M.S., et al.: Performance analysis of multiple NLOS UV communication cooperative relays over turbulent channels. Opt. Express **26**(16), 19972–19985 (2018)
23. Wang, Z., Yuan, R., Peng, M.: Non-line-of-sight full-duplex ultraviolet communications under self-interference. IEEE Trans. Wirel. Commun. **22**(11), 7775–7788 (2023)
24. Xu, Z., Chen, G., Abou-Galala, F., et al.: Experimental performance evaluation of non-line-of-sight ultraviolet communication systems. Free-Space Laser Commun. VII **6709**, 67090Y (2007)
25. Chen, G., Xu, Z., Sadler, B.M.: Experimental demonstration of ultraviolet pulse broadening in short-range non-line-of-sight communication channels. Opt. Express **18**(10), 10500–10509 (2010)
26. Meng, X., Zhang, M., Han, D., et al.: Experimental study on 1×4 real-time SIMO diversity reception scheme for a ultraviolet communication system. In: 2015 20th European Conference on Networks and Optical Communications-(NOC) pp. 1–4 (2015)
27. Sun, X., Zhang, Z., Chaaban, A., et al.: 71-Mbit/s ultraviolet-B LED communication link based on 8-QAM-OFDM modulation. Opt. Express **25**(19), 23267–23274 (2017)
28. Wang, G., Wang, K., Gong, C., et al.: A 1 Mbps real-time NLOS UV scattering communication system with receiver diversity over 1 km. IEEE Photon. J. **10**(2), 1–13 (2018)
29. Alkhazragi, O., Hu, F., Zou, P., et al.: Gbit/s ultraviolet-C diffuse-line-of-sight communication based on probabilistically shaped DMT and diversity reception. Opt. Express **28**(7), 9111–9122 (2020)

Chapter 2
Channel Modeling of Ultraviolet Communications

Abstract Different from the channel modeling in traditional optical wavebands, the modeling of ultraviolet (UV) communication channel has to take into consideration of multiple scattering effects of UV signals when traveling in the atmosphere, which brings great challenges to the modeling of UV communication channels. In this chapter, we classify the channel models of UV communication into two types, i.e., *analytical channel models* and *stochastic channel models*. This classification is based on the fact that the computational efficiency of analytical channel models is much more higher than that of stochastic channel models. We first introduce these two types of channel models in Sects. 2.1 and 2.2, respectively. Then we present some numerical results on typical channel models in Sect. 2.3. At last, we summarize this chapter and present some future directions on channel modeling of UV communications in Sect. 2.4.

Keywords Multiple-scattering models · Single-scattering models · Monte-Carlo methods

2.1 Analytical Channel Models of UV Communications

The analytical channel model characterize the scattering model as a multiple (generally not more than three) integral form that can be quickly calculated using existing computational software. Currently, analytical channel models mainly include single-scattering models, second-order scattering models, and third-order scattering models. The advantage of such models is fast calculation speed, usually in seconds, but the disadvantage is limited application scenarios, making it almost impossible to be extended to complex application scenarios. For example, the single-scattering integral model can usually only estimate channel loss within 100 m; existing second-order and third-order analytical models can only calculate channel loss under small emission angle conditions. For scattering order higher than third, at long distances above 100 m, large emission angles, or more complex scenarios (with obstacles, reflective boundaries, etc.), there is currently no suitable analytical channel model.

2.1.1 Analytical Single-Scattering Models

The single-scattering model assumes that photons are received by the receiving end after only one scattering. Figure 2.1 shows the geometric parameter definitions of the single-order scattering model. We use ϕ_1 and ϕ_2 to denote the divergence angles of the transmitted beam from the transmitter (Tx) and the field-of-view at the receiver (Rx), respectively. The linear distance between the transmitter and receiver is denoted by r, and the common volume between the beam and the FOV is denoted by V.

Suppose δV is an arbitrary differential element volume in the common volume V; r_1 is the distance between the Tx and δV; r_2 is the distance between δV and Rx; θ_1 is the angle between r_1 and r; θ_2 is the angle between r_2 and r; $\theta_s = \theta_1 + \theta_2$ is the scattering angle at δV; ζ is the angle between r_2 and the axis of the FOV. Suppose P_t is the transmitting power at Tx. Then the power arriving at δV per unit area can be expressed as

$$H_s = \frac{P_t e^{-k_e r_1}}{\Omega_t r_1^2},\tag{2.1}$$

where $\Omega_t = 2\pi \left[1 - \cos\frac{\phi_1}{2}\right]$ is the solid angle formed by the transmitting beam; $k_e = k_s + k_a$ is the extinction coefficient of the UV light, where k_s and k_a are the scattering coefficient and the absorption coefficient of the UV light, respectively.

Then the light power scattered from δV can be obtained by

$$Q_s = H_s \delta s (1 - e^{-k_s \delta r_1}) \approx \frac{P_t k_s e^{-k_e r_1}}{\Omega_t r_1^2} \delta V,\tag{2.2}$$

where δs and δr_1 are the section area and height of δV, respectively; and we have used the fact that $\delta V = \delta s \times \delta r_1$ and $1 - e^{-k_s \delta r_1} \approx k_s \delta r_1$ when $\delta r_1 \to 0$.

Then the light power per solid angle scattered to the Rx can be expressed as

$$Q_r = Q_s P(\theta_s) = \frac{P_t k_s P(\theta_s) e^{-k_e r_1}}{\Omega_t r_1^2} \delta V,\tag{2.3}$$

Fig. 2.1 Geometrical setting for single-scattering model

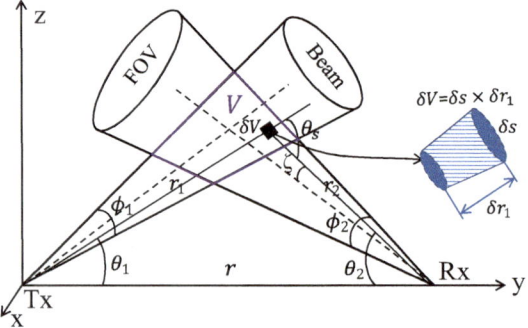

where $P(\theta_s)$ is the phase scattering function of the UV light in the atmosphere determined by both the Rayleigh scattering and Mie scattering. Specifically, we have

$$P(\theta_s) = \frac{k_s^{Ray}}{k_s} P^{Ray}(\theta_s) + \frac{k_s^{Mie}}{k_s} P^{Mie}(\theta_s), \tag{2.4}$$

where k_s^{Ray} and k_s^{Mie} are the scattering coefficients due to the Rayleigh and Mie scattering, respectively; and we have $k_s = k_s^{Ray} + k_s^{Mie}$; $P^{Ray}(\theta_s)$ and $P^{Mie}(\theta_s)$ are the phase scattering functions due to the Rayleigh and Mie scattering, respectively; and we have

$$P^{Ray}(\theta_s) = \frac{3[1 + 3\gamma + (1 - \gamma)\cos^2\theta_s]}{16\pi(1 + 2\gamma)} \tag{2.5}$$

and

$$P^{Mie}(\theta_s) = \frac{1 - g^2}{4\pi} \left[\frac{1}{(1 + g^2 - 2g\cos\theta_s)^{3/2}} + f \frac{0.5(3\cos^2\theta_s - 1)}{(1 + g^2)^{3/2}} \right], \tag{2.6}$$

where γ, f, and g are model parameters.

The phase scattering function is defined as the scattering intensity per solid angle. Therefore, it satisfies the following normalization condition:

$$\int_\Omega P(\theta_s)\mathrm{d}\Omega = \int_0^\pi \int_0^{2\pi} P(\theta_s)\sin\theta_s\mathrm{d}\theta_s\mathrm{d}\phi_s = 1. \tag{2.7}$$

For example, when the wavelength of the UV light is $\lambda = 265$ nm, we have $k_s^{Ray} = 0.266 \times 10^{-3}$ km^{-1}, $k_s^{Mie} = 0.284 \times 10^{-3}$ km^{-1}, $k_a = 0.802 \times 10^{-3}$ km^{-1}, $\gamma = 0.017$, $f = 0.5$, and $g = 0.72$. The phase scattering functions at various scattering angles are given in Fig. 2.2. From Fig. 2.2 we can see that the Rayleigh scattering effect approximates an isotropic scattering; while the Mie scattering effect is dominated by the forward scattering.

Fig. 2.2 Phase scattering function for wavelength 265 nm in the atmosphere

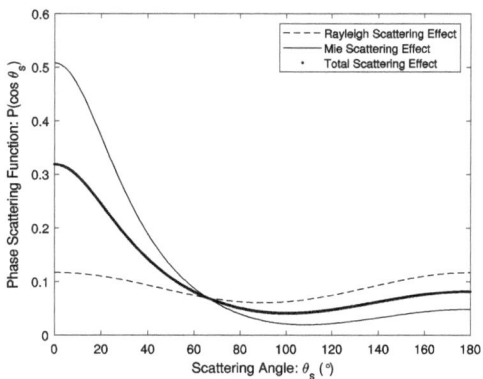

Then the receiving power at the Rx due to the scatter δV can be expressed as

$$\delta P_{r,1} = \Omega_r Q_r e^{-k_e r_2} = \frac{A_r \cos \zeta}{r_2^2} Q_r e^{-k_e r_2}, \tag{2.8}$$

where $\Omega_r = \frac{A_r \cos \zeta}{r_2^2}$ is the solid angle formed by the receiving aperture and the scatter δV and A_r is the receiving area of the receiving aperture at Rx. Substituting Eq. (2.3) into (2.8), we can obtain

$$\delta P_{r,1} = \frac{P_t A_r k_s P(\theta_s) \cos \zeta}{\Omega_t r_1^2 r_2^2} e^{-k_e(r_1+r_2)} \delta V. \tag{2.9}$$

Finally, the receiving power at Rx due to the single-scattering effect happened in the common volume V can be expressed as

$$P_{r,1} = \int_V \delta P_{r,1} = \int_V \frac{P_t A_r k_s P(\theta_s) \cos \zeta}{\Omega_t r_1^2 r_2^2} e^{-k_e(r_1+r_2)} \delta V. \tag{2.10}$$

From Eq. (2.10) we can see that the receiving power $P_{r,1}$ highly depends on the shape of the common volume V formed by the intersection between the transmitting beam and the FOV. Using different coordinate systems, we can express the $P_{r,1}$ as a triple integral in different forms. For example, the first accurate single-scattering model for calculating $P_{r,1}$ in a coplanar geometry was first established by Reilly based on a prolate-spheroidal coordinates system, and was improved by Luettgen in 1991 [1]. In 2011, Elshimy et al. [2] derived an accurate single-scattering model for any transmitting and receiving geometry based on an prolate-spheroidal coordinate system.

Specifically, the prolate-spheroidal coordinates system uses three variables ξ, η, and ϕ to describe a point in the common volume V, where ξ, η, and ϕ are given by

$$\begin{cases} \xi = \frac{r_1+r_2}{r}, \\ \eta = \frac{r_1-r_2}{r}, \\ \phi = \arctan \frac{z}{x}, \end{cases} \tag{2.11}$$

where (x, y, z) is the coordinates of δV in a $O - x - y - z$ coordinate system as shown in Fig. 2.3.

For a given ξ, we have $r_1 + r_2 = \xi r$, which is a constant number. Therefore, the receiving time for a given ξ is given by

$$t = \frac{\xi r}{c}, \tag{2.12}$$

Fig. 2.3 Geometrical setting for the single-scattering model in a prolate-spheroidal coordinates system

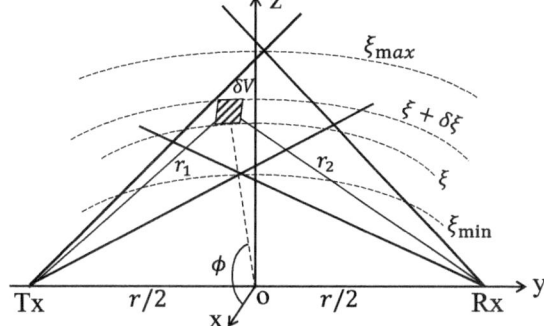

where c is the light speed. In this context, the prolate-spheroidal coordinate system can conveniently calculate the channel impulse response (CIR) of the UV communication channel. Specifically, we can express the CIR for the single-scattering model as

$$h_1(t) = \frac{ck_s A_r \exp(-k_e ct)}{4\pi^2(1 - \cos\frac{\beta_T}{2})r^2} \int_{\eta_1(t)}^{\eta_2(t)} \int_{\phi_1(t,\eta)}^{\phi_2(t,\eta)} \frac{\cos(\zeta)P(\theta_s)}{\left(\frac{ct}{r}\right)^2 - \eta^2} d\phi d\eta, \qquad (2.13)$$

where $\eta_1(t)$ and $\eta_2(t)$ are integral limits for η given the time instant t; and $\phi_1(t, \eta)$ and $\phi_2(t, \eta)$ are integral limits for ϕ given both the time instant t and η. The integral limits $\eta_1(t)$, $\eta_2(t)$, $\phi_1(t, \eta)$, and $\phi_2(t, \eta)$ are determined by the shape of the common volume V. The explicit expressions for the integral limits $\eta_1(t)$, $\eta_2(t)$, $\phi_1(t, \eta)$, and $\phi_2(t, \eta)$ are complex and closely related to the geometries of the transmitter and the receiver. In 2012, Zuo et al. [3] derived an accurate single-scattering model for any transmitting and receiving geometry based on a spherical coordinate system. The spherical coordinate systems does not require complex coordinate transformations, but spherical coordinate systems are not convenient for calculating the CIR.

2.1.2 Approximated Single-Scattering Models

In addition to accurate single-scattering models, various approximated single-scattering models have been proposed to simplify the computational difficulty [16, 19]. For example, when the transmitting divergence angle ϕ_1 is small enough and the FOV angle is relatively small, we can approximate V as a frustum of beam cone shown in Fig. 2.4 with its volume given by

$$V = \frac{\pi(D_1^2 h_1 - D_2^2 h_2)}{3}, \qquad (2.14)$$

where $h_1 = r_1 + r_2\phi_2/2$ and $D_1 = h_1\phi_1/2$ are the height and the radius of the top surface of the frustum; $h_2 = r_1 - r_2\phi_2/2$ and $D_2 = h_2\phi_1/2$ are the height and the

Fig. 2.4 Frustum
approximation of the
common volume

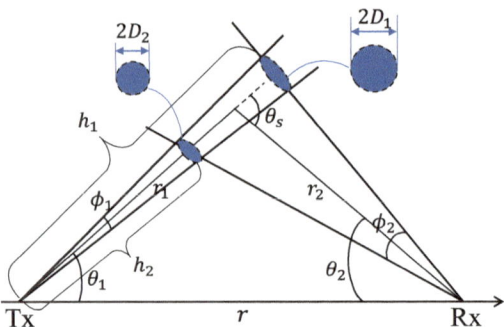

radius of the bottom surface of the frustum. By supposing that the scattering angle θ_s in V are close to each other, we can approximate the receiving power of the single-scattering model as

$$P_{r,1} = \frac{P_t k_s P(\theta_s) A_r \phi_1^2 \phi_2 \sin \theta_s \left(12 \sin^2 \theta_2 + \phi_2^2 \sin^2 \theta_1\right)}{96r \sin \theta_1 \sin^2 \theta_2 \left(1 - \cos \frac{\phi_1}{2}\right)} e^{-\frac{k_e r (\sin \theta_1 + \sin \theta_2)}{\sin \theta_s}}. \qquad (2.15)$$

Equation (2.15) provides a closed-form expression for the receiving power of single-scattering model, which is useful in performance analysis of either NLOS UV communication or NLOS UV positioning.

There are many other approximations on the common volume, resulting in different approximated single-scattering models. For examples, in 2010, Yin et al. derived the formula for calculating the receiving power when the scattering in V is isotropic [4]; in 2011, Wang et al. derived an approximated closed-form expression in an non-coplanar geometry [5]; in 2019, Wu et al. derived the receiving power at narrow beam angles using the spherical crown approximation method [6].

2.1.3 Analytical Multiple-Scattering Models

The single-scattering model only validates in short-range UV communications. As the communication range increases, multiple scattering effects becomes non-negligible. Here we consider the analytical model for the second-order scattering and the third-order scattering only.

The second-order scattering model assumes that photons are scattered twice to reach the receiving end. Figure 2.5 shows the geometrical setting for the second-order scattering model. We denote the zenith and azimuth angles for the Tx (Rx) by $\theta_{t(r)}$ and $\phi_{t(r)}$, respectively. The beam divergence angle is β_t and the FOV angle is β_r. The distance between Tx and Rx is r.

Suppose δv_1 and δv_2 are the first and the second scatter. We use bold symbols to denote vectors and its corresponding non-bold symbols to denote the norm of

Fig. 2.5 Geometrical setting for the second-order scattering model

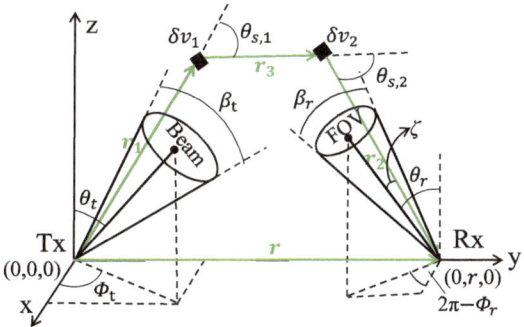

vectors. The vector from Tx to δv_1 is r_1; the vector from δv_2 to Rx is r_2; the vector from δv_1 to δv_2 is r_3. Let P_t be the transmitting power. According to the derivation of the single-scattering model, we can express the light power scattered from δv_1 as

$$Q_{s,1} = \frac{P_t k_s e^{-k_e r_1}}{\Omega_t r_1^2} \delta v_1, \qquad (2.16)$$

where $\Omega_t = 2\pi(1 - \cos(\beta_t/2))$ is the solid angle of the transmitting beam.

By regarding δv_1 as a new light source with transmitting power $Q_{s,1}$, then using similar derivation process, we can obtain the light power scattered from δv_2 as

$$Q_{s,2} = \frac{Q_{s,1} k_s P(\theta_{s,1}) e^{-k_e r_3}}{r_3^2} \delta v_2, \qquad (2.17)$$

where $\theta_{s,1}$ is the scattering angle at δv_1, i.e., the angle between r_1 and r_3.

Then the receiving power due to the second-order scattering happened at δv_1 and δv_2 can be obtained as

$$\delta P_{r,2} = Q_{s,2} \frac{A_r \cos \zeta P(\theta_{s,2}) e^{-k_e r_2}}{r_2^2}, \qquad (2.18)$$

where $\theta_{s,2}$ is the scattering angle at δv_2, i.e., the angle between r_3 and r_2. By substituting (2.16) and (2.17) into (2.18), we can further obtain

$$\delta P_{r,2} = P_t \frac{A_r k_s^2 P(\theta_{s,1}) P(\theta_{s,2}) \cos \zeta e^{-k_e(r_1+r_2+r_r)}}{\Omega_t r_1^2 r_2^2 r_3^2} \delta v_1 \delta v_2. \qquad (2.19)$$

Therefore, the total receiving power of the second-order scattering model can be obtained as

$$P_{r,2} = \int_{V_1} \int_{V_2} P_t \frac{A_r k_s^2 P(\theta_{s,1}) P(\theta_{s,2}) \cos \zeta e^{-k_e(r_1+r_2+r_3)}}{\Omega_t r_1^2 r_2^2 r_3^2} \delta v_1 \delta v_2, \qquad (2.20)$$

Fig. 2.6 Geometrical setting
for the third-order scattering
model

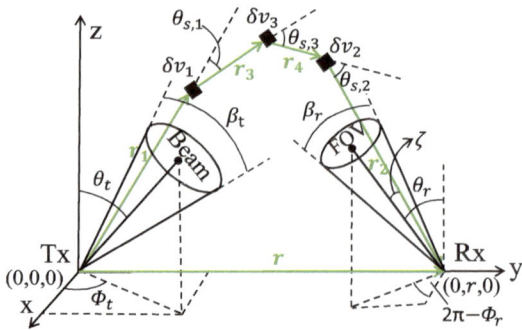

where V_1 and V_2 are the integral areas for δv_1 and δv_2, respectively. Specifically, V_1 is the whole transmitting beam and V_2 is the whole FOV cone.

Similarly, the third-order scattering model is based on the assumption that a photon is scattered three times before arriving at Rx. We set the geometrical parameters for a third-order scattering model in Fig. 2.6. By using the same derivation method, we can obtain the receiving power due to the third-order scattering as

$$P_{r,3} = \int_{V_1} \int_{V_2} \int_{V_3} P_t \frac{A_r k_s^3 P(\theta_{s,1}) P(\theta_{s,2}) P(\theta_{s,3}) \cos\zeta e^{-k_e(r_1+r_2+r_3+r_4)}}{\Omega_t r_1^2 r_2^2 r_3^2 r_4^2} \delta v_1 \delta v_2 \delta v_3,$$

$$(2.21)$$

where V_1 is the whole transmitting beam, V_2 is the whole FOV cone, and V_3 is the whole space.

Unfortunately, the integrals in (2.20) and (2.21) are respectively six-fold integral and nine-fold integral, which have no analytical solutions in practical implementations. In 2016, Yuan et al. proposed an approximated second-order scattering model and an approximated third-order scattering model by considering narrow beam and narrow FOV cases [7]. Here we introduce them in the following subsection.

2.1.4 Approximated Multiple-Scattering Models

2.1.4.1 Approximated Second-Order Scattering Model

When the transmitting beam and FOV are narrow, we can approximate δv_1 and δv_2 as two thin layers of spherical caps, as shown in Fig. 2.7. Then we can express δv_1 and δv_2 as

$$\begin{cases} \delta v_1 = \delta s_1 \delta r_1 = 2\pi r_1^2 (1 - \cos\frac{\beta_t}{2}) \delta r_1, \\ \delta v_2 = \delta s_2 \delta r_2 = 2\pi r_2^2 (1 - \cos\frac{\beta_t}{2}) \delta r_2. \end{cases} \qquad (2.22)$$

Fig. 2.7 Geometrical setting for the second-order scattering model in narrow beam and FOV cases

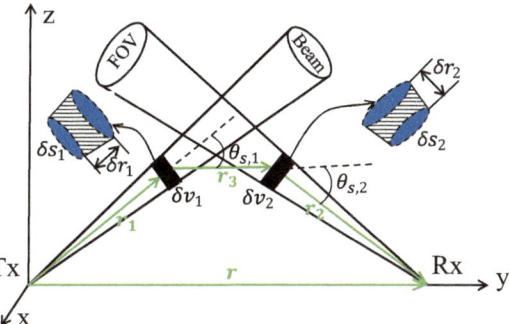

Noting that when FOV is narrow, we have $\cos \zeta = 1$. Then we can approximate (2.20) as

$$P_{r,2} \approx \int_0^\infty \int_0^\infty P_t \frac{A_r \Omega_{FOV} k_s^2 P(\theta_{s,1}) P(\theta_{s,2}) e^{-k_e(r_1+r_2+r_3)}}{r_3^2} \delta r_1 \delta r_2, \qquad (2.23)$$

where $\Omega_{FOV} = 2\pi(1 - \cos \frac{\beta_r}{2})$ is the solid angle of the FOV cone.

2.1.4.2 Approximated Third-Order Scattering Model

For the third-order scattering model in narrow beam and FOV cases, we can also approximate the receiving power in (2.21) as

$$P_{r,3} \approx \int_0^\infty \int_0^\infty M P_t A_r \Omega_{FOV} k_s^3 e^{-k_e(r_1+r_2)} \delta r_1 \delta r_2, \qquad (2.24)$$

where M is defined as

$$M \triangleq \int_{V_3} \frac{P(\theta_{s,1}) P(\theta_{s,2}) P(\theta_{s,3}) e^{-k_e(r_3+r_4)}}{r_3^2 r_4^2} \delta v_3. \qquad (2.25)$$

To estimate M, we assume that the scattering effect is isotropic, i.e., $P(\theta_{s,1}) = P(\theta_{s,2}) = P(\theta_{s,3}) = \frac{1}{4\pi}$. Besides, we further use the two scatters δv_1 and δv_2 as two focuses to form a prolate-spheroidal coordinate system, as shown in Fig. 2.8. A point in the prolate-spheroidal coordinate system is described as

$$\begin{cases} \xi = \frac{r_3+r_4}{r_5}, \\ \eta = \frac{r_3-r_4}{r_5}, \\ \phi = \arctan \frac{z'}{x'}, \end{cases} \qquad (2.26)$$

Fig. 2.8 Geometrical setting for the third-order scattering model in narrow beam and FOV cases

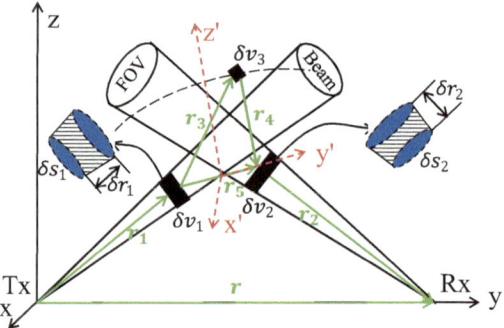

where r_5 is the distance between δv_1 and δv_2. In the prolate-spheroidal coordinate system, we have

$$\delta v_3 = \frac{r_5^3}{8}(\xi^2 - \eta^2)\delta\xi\delta\eta\delta\phi. \tag{2.27}$$

Then we can obtain M as

$$M = \int_\xi \int_\eta \int_\phi \frac{2e^{-k_e r_5\xi}}{(4\pi)^3 r_5(\xi^2 - \eta^2)}\delta\xi\delta\eta\delta\phi. \tag{2.28}$$

Because V_3 is the whole space, the integral limits for ξ, η, and ϕ are $[1, \infty]$, $[-1, 1]$, and $[0, 2\pi]$, respectively. Then we can further obtain M as

$$M = \int_1^\infty \frac{e^{-k_e r_5\xi}}{16\pi^2 r_5\xi} \ln\left(\frac{\xi + 1}{\xi - 1}\right)\delta\xi. \tag{2.29}$$

Then the receiving power due to the third-order scattering can be approximated as

$$P_{r,3} \approx \int_0^\infty \int_0^\infty \int_1^\infty P_t \frac{A_r\Omega_{FOV}k_s^3 e^{-k_e(r_1+r_2+r_5\xi)}}{16\pi^2 r_5\xi} \ln\left(\frac{\xi + 1}{\xi - 1}\right)\delta r_1\delta r_2\delta\xi. \tag{2.30}$$

2.2 Stochastic Channel Models of UV Communications

Stochastic channel models simulate the emission, transmission, scattering, and reception processes of photons based on probability theory, and use Monte-Carlo methods to calculate the receiving probability of photons. Current stochastic channel models are mainly divided into Monte-Carlo simulation (MCS) models [8, 9] and Monte-Carlo integration (MCI) models [10, 11]. The advantages of stochastic channel models are that it can model long-distance, arbitrary transmitting and receiving geometry, and complex scenes and it can calculate the CIR conveniently. However, the disadvantage of stochastic channel models is their low computational efficiency, as Monte-Carlo methods require averaging the simulation results of a large number of photons (usually above the order of 10^6). Therefore, how to improve the computational efficiency of stochastic channel models has always been a research focus and difficulty for NLOS UV communications.

2.2.1 Monte-Carlo Simulation Model

The Monte-Carlo simulation (MCS) model starts from the perspective of mathematical statistics and establishes corresponding probability models or random processes for the emission, transmission, scattering, and reception of photons. Then, a large number of photons are used to repeat the entire random process, calculate the probability of receiving each photon, and calculate the corresponding receiving time. The former can be used to calculate the path loss of UV communications, while the latter can be used to calculate the CIR of UV communications.

2.2.1.1 Stochastic Channel Modeling

Figure 2.9 shows the geometric parameter definition of the MCS model. For the convenience of calculating the receiving probability, the receiver (Rx) is placed at the coordinate origin $(0, 0, 0)$, and the transmitter (Tx) is located at the coordinate $(0, r, 0)$. The transmitting direction is characterized by the beam zenith angle θ_t and beam azimuth angle ϕ_t; the receiving direction is characterized by the FOV zenith angle θ_r and FOV azimuth angle ϕ_r. The beam divergence angle is denoted by β_t and the FOV angle is denoted by β_r. We denote the direction cosine of the transmitting beam by $\boldsymbol{\mu}_t$ and the direction cosine of the receiving FOV axis by $\boldsymbol{\mu}_r$, where $\boldsymbol{\mu}_t$ and $\boldsymbol{\mu}_r$ are respectively given by

Fig. 2.9 Geometrical setting for the third-order scattering model in narrow beam and FOV cases

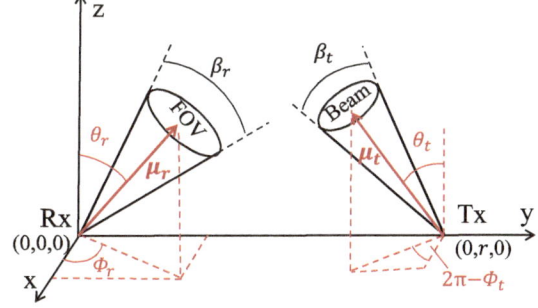

$$\begin{cases} \boldsymbol{\mu}_t = [\cos \phi_t \sin \theta_t, \sin \phi_t \sin \theta_t, \cos \theta_t]^{\mathrm{T}}, \\ \boldsymbol{\mu}_r = [\cos \phi_r \sin \theta_r, \sin \phi_r \sin \theta_r, \cos \theta_r]^{\mathrm{T}}, \end{cases} \tag{2.31}$$

where $\boldsymbol{x}^{\mathrm{T}}$ denotes the transpose of vector $\boldsymbol{x}$.

Considering a n-order scattering process. Let (θ_0, ϕ_0) be the zenith and azimuth angles of the photon emitting direction angles from the Tx. Assuming that the light source is uniformly distributed, then the probability density functions (PDFs) of θ_0 and ϕ_0 are given by

$$\begin{cases} f(\theta_0) = \frac{\sin(\theta_0)}{1 - \cos \frac{\beta_t}{2}}, \\ f(\phi_0) = \frac{1}{2\pi}. \end{cases} \tag{2.32}$$

Then the initial emission direction of a photon can be generated by the following equations:

$$\begin{cases} \theta_0 = \arccos \left(1 - \text{rand}(1)(1 - \cos \frac{\beta_t}{2}) \right), \\ \phi_0 = 2\pi \cdot \text{rand}(1), \end{cases} \tag{2.33}$$

where rand(1) randomly generate a real number ranging from [0, 1].

The propagating distance d of a photon before it is scattered when departing from the Tx or a scatterer satisfies an exponential distribution with PDF given by:

$$f(d) = k_s e^{-k_s d}, \tag{2.34}$$

where k_s is the scattering coefficient of the UV light. Then the propagating distance of the photon after leaving the ith scatterer (where i=0 corresponds to the photon leaving the emitting end) is generated by the following equation:

$$d_i = -\frac{\ln(1 - \text{rand}(1))}{k_s}. \tag{2.35}$$

The direction of photons after being scattered by scatterers changes. let (θ_i, ϕ_i) be the zenith angle and azimuth angle at which the photon deviates from its original direction after the ith scattering, where $i = 1, 2, \ldots, n$. According to the phase scattering function, the PDFs of θ_i and ϕ_i are given by

$$\begin{cases} f(\theta_i) = 2\pi P(\theta_i) \sin \theta_i, \\ f(\phi_i) = \frac{1}{2\pi} \end{cases} \tag{2.36}$$

where $P(\theta_i)$ is the phase scattering function given in (2.4).

Then (θ_i, ϕ_i) can be generated by the following equations:

$$\begin{cases} \theta_i = F^{-1}(\text{rand}(1)), \\ \phi_i = 2\pi \cdot \text{rand}(1), \end{cases} \tag{2.37}$$

where $F(\theta_i)$ is the marginal distribution of θ_i over $f(\theta_i, \phi_i)$; and $F(\theta_i)$ is given by

$$F(\theta_i) = \int_0^{\theta_i} 2\pi P(\theta_i) \sin \theta_i d\theta_i. \tag{2.38}$$

If a photon is located within the receiving field of view (FOV) after each scattering, the receiving probability that a photon arrives at detector can be calculated. Assuming that the solid angle formed between the ith scatter and the receiving aperture is Ω_i, then Ω_i can be approximated as

$$\Omega_i \approx \frac{A_r \cos \zeta}{r_i^2}, \tag{2.39}$$

where $\cos \zeta = \frac{-\mu_r \cdot r_i}{r_i}$, and where r_i is the vector of the ith scatter.

We denote the angle between the original photon propagating direction and r_i by θ_{r_i}; then the photon receiving probability after it is scattered by the ith scatter is given by [8]:

$$P_{r,i} = \begin{cases} p_{s,i-1}e^{-k_a d_{i-1}} \min(1, P(\theta_{r_i})\Omega_i)e^{-k_e r_i}, & r_i \text{ locates in FOV,} \\ 0, & \text{otherwise,} \end{cases} \tag{2.40}$$

where k_e is the extinction coefficient of the UV light; $p_{s,i-1}$ is the photon survival probability before it arrives at the ith scatter; and the photon survival probability needs to be updated after each receiving process as

$$p_{s,i} = p_{s,i-1}e^{-k_a d_{i-1}}(1 - \min(1, P(\theta_{r_i})\Omega_i)), \tag{2.41}$$

where the min function here is to ensure that the probability $\min(1, P(\theta_{r_i})\Omega_i)$ cannot exceed 1.

2.2.1.2 Monte-Carlo Simulation Process

The MCS process for one photon is shown in Fig. 2.10. A photon is first generated and the scattering order $i = 0$ and survival probability $p_{s,0} = 1$ are initialized. Then we increase the scattering order $i = i + 1$ and judge if the scattering order exceed the maximum scattering order n or the survival probability is smaller than a pre-chosen minimum survival probability $p_{s,min}$. If yes, then the process for current photon is ended and we generate a new photon if current photon is not the last photon; if no, we generate the photon scattering parameters θ_{i-1} and ϕ_{i-1} according to (2.33) or (2.37) and generate the propagating distance d_{i-1} according to (2.35). Then we judge if the photon passes through the receiving area. If yes, we have to regenerate the photon scattering parameters and propagating distance; if no, we update the photon location and survival probability. Then we compute the receiving probability and receiving

Fig. 2.10 Monte-Carlo
simulation process for one
photon

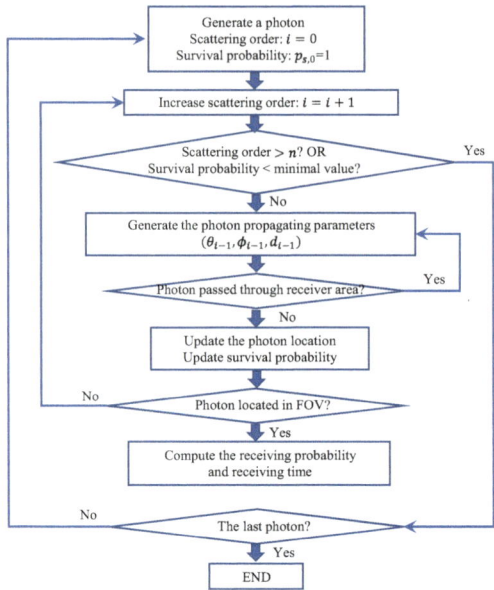

time if the photon locates in FOV, otherwise we increase the scattering order and
repeat above processes.

In 2011, Drost et al. [9] proposed an improved MCS model by omitting the
survival probability and improving the sampling efficiency of the distance by using
the following PDF:

$$f(d) = k_e e^{-k_e d}. \tag{2.42}$$

It is noted that this distance sampling method defined in (2.42) of [9] is equivalent to the original distance sampling method defined in (2.34) of [8], where the
former is based on both the scattering and absorption effects and the later is based
on the scattering effect only and the absorption effect is regarded as a energy loss in
propagating process.

2.2.2   Monte-Carlo Integration Model

The Monte-Carlo integration (MCI) model is a probability integration method, which
first represents the receiving power as a probability integration on photon transmission parameters and then uses Monte-Carlo integration techniques to obtain this
probability integration [11]. The MCI model has been proven to be able to calculate
the CIR for NLOS UV communications. Compared with the MCS model, the computational structure is simpler and the computational efficiency of MCI models can
be optimized by adopting different sampling functions.

2.2.2.1 Receiving Probability of n-Order Scattering

Figure 2.11 shows the parameter definition of the multiple scattering process for MCI model. The direction cosine vectors μ_t and μ_r are defined the same as those in Fig. 2.9. Assuming that the initial photon emission process is the zero-order scattering process, we can use d_i, θ_i, and ϕ_i to denote the photon transmission distance, scattering zenith angle, and scattering azimuth angle for the ith order scattering, respectively. The vector of the ith scatter is denoted by r_i, and the direction vector of the photon after the ith scattering is denoted by μ_i.

For the ith scattering, the probability that a photon after scattering locates in a differential solid angle $d\Omega_i = \sin\theta_i d\theta_i d\phi_i$ and the propagating distance is in range $(d_i, d_i + dd_i)$ can be expressed as

$$dQ_i = f(d_i)f(\theta_i)f(\phi_i)d\theta_i d\phi_i dd_i, \tag{2.43}$$

where $f(d_i)$ is the PDF of propagating distance, and here we adopt the PDF in (2.42) to improve the sampling efficiency; $f(\phi_i) = \frac{1}{2\pi}$ is the PDF of scattering azimuth angle; and $f(\theta_i)$ is the PDF of the scattering zenith angle. Specifically, when $i = 0$, $f(\theta_i)$ is given in (2.32); and when $i > 0$, $f(\theta_i)$ is given in (2.36).

Then the photon receiving probability after it is scattered by the nth scatter is given by [11]

$$p_d = \begin{cases} \frac{k_s}{k_e}e^{-k_e d_n}\min(1, P(\theta_n)\Omega_n), & r_n \text{ locates in FOV}, \\ 0, & \text{otherwise}, \end{cases} \tag{2.44}$$

where θ_n is the angle between the photon propagating direction μ_{n-1} and $-r_n$; $\Omega_n \approx \frac{A_r \cos\zeta}{r_n^2}$ is the solid angle formed by the receiving aperture and the nth scatter.

Then we can express the receiving probability of a photon arrives at the Rx after n-order scattering as

Fig. 2.11 Geometry setting for Monte-Carlo integration process

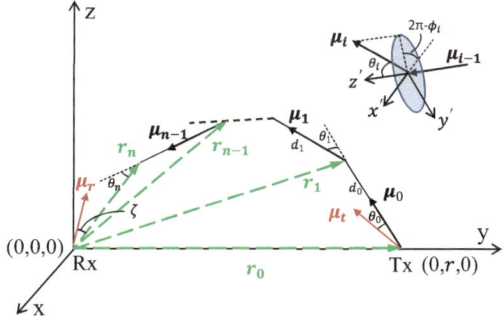

$$P_n = \int_V p_d \left(\frac{k_s}{k_e}\right)^{n-1} \mathrm{d}Q_0 \cdots \mathrm{d}Q_{n-1}$$

$$= \int_V p_d \left(\frac{k_s}{k_e}\right)^{n-1} \left[\prod_{i=0}^{n-1} f(d_i) f(\theta_i) f(\phi_i) \mathrm{d}d_i \mathrm{d}\theta_i \mathrm{d}\phi_i\right], \qquad (2.45)$$

where V is a $3n$-dimensional integral area.

2.2.2.2 Monte-Carlo Integration Method

We have expressed the receiving probability of n-order scattering as a $3n$-fold integral in (2.45). Now we can resort to the widely used Monte-Carlo integration method to estimate this $3n$-fold integral. Specifically, we first rewrite (2.45) as

$$P_n = \int_V g_n \left[\prod_{i=0}^{n-1} \mathrm{d}d_i \mathrm{d}\theta_i \mathrm{d}\phi_i\right], \qquad (2.46)$$

where g_n is the integrand defined as

$$g_n \triangleq p_d \left(\frac{k_s}{k_e}\right)^{n-1} \left[\prod_{i=0}^{n-1} f(d_i) f(\theta_i) f(\phi_i)\right]. \qquad (2.47)$$

Then we choose a sampling PDF f_n defined on the integral space V and introduce an objective function O_n^* defined as

$$O_n^* \triangleq \frac{g_n(\theta_0, \phi_0, d_0, \ldots, \theta_{n-1}, \phi_{n-1}, d_{n-1})}{f_n(\theta_0, \phi_0, d_0, \ldots, \theta_{n-1}, \phi_{n-1}, d_{n-1})}. \qquad (2.48)$$

Then we can further rewrite P_n as

$$P_n = \int_V O_n^* f_n \left[\prod_{i=0}^{n-1} \mathrm{d}d_i \mathrm{d}\theta_i \mathrm{d}\phi_i\right] = \mathrm{E}[O_n^*], \qquad (2.49)$$

which means that the receiving power P_n is the expectation of the objective function O_n^* under sampling PDF f_n.

Suppose we have randomly generated N sampling points $x_1, x_2, \ldots, x_N$ with $x_i \triangleq (\theta_{i,0}, \phi_{i,0}, d_{i,0}, \ldots, \theta_{i,n-1}, \phi_{i,n-1}, d_{i,n-1})$ is a $3n$-dimensional vector. Then according to the law of large numbers, we can estimate the expectation of O_n^* by

$$P_n = \lim_{N \to \infty} \frac{1}{N} \sum_{i=1}^{N} O_n^*(x_i). \qquad (2.50)$$

Besides, sort all $O_n^*(x_i)$ into receiving time slot $(T_1, T_2, \ldots, T_m)$ and divided all receiving probability in a time slot with total sampling numbers N, we can obtain the following sequence of receiving probabilities:

$$\{P_{n,T_1}, P_{n,T_2}, \ldots, P_{n,T_m}\}. \tag{2.51}$$

Normalize above sequence with time interval ΔT and receiving area A_r, we can obtain the CIR in units of irradiance.

2.2.2.3 Three Sampling Methods

Because the sampling function f_n can be arbitrarily selected and a different f_n can result in different $\sigma(O_n^*)$, the convergence performance of an MCI model mainly depends on the selection of f_n given N sampling points. In fact, the more similar the shapes of functions f_n and g_n are, the faster the convergence speed is. An importance sampling is achieved when $f_n = g_n/P_n$, and its convergence speed is the fastest [12]. Because p_d is decided by the geometric parameters $(d_0, \theta_0, \phi_0, \ldots, d_{n-1}, \theta_{n-1}, \phi_{n-1})$ of the previous $n-1$ scattering orders, once sampling processes for the previous $n-1$ scattering orders are completed, the value of p_d can be calculated correspondingly. Therefore, we only need to consider the sampling of variables $(d_0, \theta_0, \phi_0, \ldots, d_{n-1}, \theta_{n-1}, \phi_{n-1})$ in the previous $n-1$ scattering orders.

Given the statistical independence of all the random variables in g_n and the fact that there are only three types of variables (propagating distance d, scattering zenith angle θ and scattering azimuth angle ϕ), sampling functions for these three types of variables can be considered separately. The importance sampling for ϕ is a uniform sampling, i.e., the sampling speed for uniform sampling is the fastest in all sampling methods; therefore, the uniform sampling for ϕ is the most efficient choice. Then the main problem here is to choose sampling functions for the propagating distance d and the scattering zenith angle θ. Therefore, we can consider the following three sampling methods in the MCI approach: uniform sampling, importance sampling, and partial importance sampling, which respectively refer to the MCI-US model, the MCI-IS model, and the MCI-PIS model.

2.2.2.4 MCI-US Model

If both scattering zenith angle θ and propagating distance d are sampled according to uniform PDF, the final sampling function f_n becomes a uniform PDF in the integral volume V

$$f_n = \frac{1}{V} = \Pi_{i=0}^{n-1}\left[\frac{1}{d_{\max}}\frac{1}{2\pi}\right]\Pi_{i=1}^{n-1}\left[\frac{1}{\pi}\right]\frac{2}{\beta_T} \tag{2.52}$$

where $d_{\max}$ is the maximum propagating distance setting for the sampling distance. Normally we can choose $5r \leq d_{max} \leq 10r$, where r is the distance between the transmitter and the receiver, because the probability that the propagating distance is greater than $5r$ is negligible.

Accordingly, the propagating distance is sampled by

$$d_i = \text{rand}(1) \cdot d_{max}, \quad i = 0, 1, \ldots, n-1 \tag{2.53}$$

where rand(1) is a random number between 0 and 1.

The scattering zenith angle is sampled by

$$\theta_i = \begin{cases} \text{rand}(1) \cdot \frac{\beta_T}{2}, \quad i = 0 \\ \text{rand}(1) \cdot \pi, \quad i = 1, 2, \ldots, n-1. \end{cases} \tag{2.54}$$

The scattering azimuth angle is sampled by

$$\phi_i = \text{rand}(1) \cdot 2\pi, \quad i = 0, 1, \ldots, n-1. \tag{2.55}$$

Then the objective function can be calculated by

$$O_n^* = g_n \frac{\beta_T}{2} \Pi_{i=0}^{n-1} [2\pi d_{\max}] \Pi_{i=1}^{n-1} \pi. \tag{2.56}$$

The uniform sampling is the simplest sampling method due to its easy implementation. It was adopted in the first MCI model proposed in [10]. However, uniform sampling normally suffers poor convergence performance.

2.2.2.5 MCI-IS Model

If both scattering zenith angle θ and propagating distance d are sampled according to their PDFs, the sampling function f_n becomes an importance sampling function

$$f_n = \Pi_{i=0}^{n-1} [f_D(d_i) f_\Theta(\theta_i) f_\Phi(\phi_i)]. \tag{2.57}$$

Accordingly, the propagating distance is sampled by

$$d_i = -\frac{\ln(1 - \text{rand}(1))}{k_e}, \quad i = 0, 1, \ldots, n-1. \tag{2.58}$$

The scattering zenith angle for the initial emitting light is sampled by

$$\theta_0 = \arccos \left[1 - \text{rand}(1) \cdot (1 - \cos \frac{\beta_T}{2}) \right] \tag{2.59}$$

and the other scattering zenith angles are sampled by

$$\theta_i = F^{-1}(\text{rand}(1)), \; i = 1, 2, \ldots, n-1 \tag{2.60}$$

where $F(\theta_i)$ is the corresponding cumulative distribution function (CDF) for the PDF of scattering zenith angle $f(\theta_i)$, and $F^{-1}(\cdot)$ is the inverse function of $F(\theta_i)$. The scattering azimuth angle is sampled by (2.55).

Then the objective function becomes

$$O_n^* = p_d \left(\frac{k_s}{k_e} \right)^{n-1}. \tag{2.61}$$

Theoretically, the importance sampling has the best convergence performance; however, it increases the complexity of the MCI model. Because the MCS model in [9] adopts the same importance sampling method for all the random variables as an MCI-IS model, we can interpret the MCS model as a special case of MCI model using importance sampling.

2.2.2.6 MCI-PIS Model

If one of the two random variables (d and θ) adopts uniform sampling and the other adopts importance sampling, then the final sampling function becomes a partial importance sampling function. Because the receiving probability for n-order scattering mainly depends on the total propagating distance of the photon, the main factor affecting convergence performance is the sampling method for propagating distance d. The phase function scattering zenith angle θ greatly increases the complexity of the MCI process. Therefore, we adopt the importance sampling for propagating distance d and the uniform sampling for scattering zenith angle θ. Now the sampling function f_n becomes

$$f_n = \Pi_{i=0}^{n-1} [f_D(d_i) f_\Phi(\phi_i)] \, \Pi_{i=1}^{n-1} \left[\frac{1}{\pi} \right] \frac{2}{\beta_T}. \tag{2.62}$$

Accordingly, the propagating distance is sampled by (2.58); the scattering zenith angle is sampled by (2.54); the scattering azimuth angle is sampled by (2.55).

Then the objective function becomes

$$O_n^* = p_d \left(\frac{k_s}{k_e} \right)^{n-1} \Pi_{i=0}^{n-1} [f_\Theta(\theta_i)] \frac{\beta_T}{2} \Pi_{i=1}^{n-1} \pi. \tag{2.63}$$

A partial importance sampling may lose some convergence performance compared to an importance sampling. However, if the calculation complexity can be decreased greatly, it will have a better computation efficiency.

2.2.2.7 Monte-Carlo Integration Process

The entire MCI process is shown in Fig. 2.12. Before implementing an MCI process, we set the maximum scattering order n and the maximum number of simulation photons N as the halt conditions. Let $i = 0$ and $k = 1$ be the initialized scattering order and the photon number, respectively. For each photon, we first sample the propagating distance d, scattering zenith angle θ, and scattering azimuth angle ϕ of each scattering order i according to the sampling function f_n. Then we update the photon location and compute the objective function O_n^* and the receiving time if the photon is located within FOV area.

Compared with the MCS process given in Fig. 2.10, we can observe that an intermediate variable called survival probability is introduced in the MCS process and a geometrical restriction (the photon may not pass through the receiving area between two scattering centers) is added. These extra calculations for survival probability and geometrical restriction make the MCS process more complicated than the MCI process. The geometry restriction in the MCS model is meaningful due to the completeness of the physical process of a photon travelling in a physical space. The sampling process for every variable in the MCI model is considered in a mathematical sampling space; therefore, this geometry restriction can be removed.

Fig. 2.12 Monte-Carlo integration process for one photon

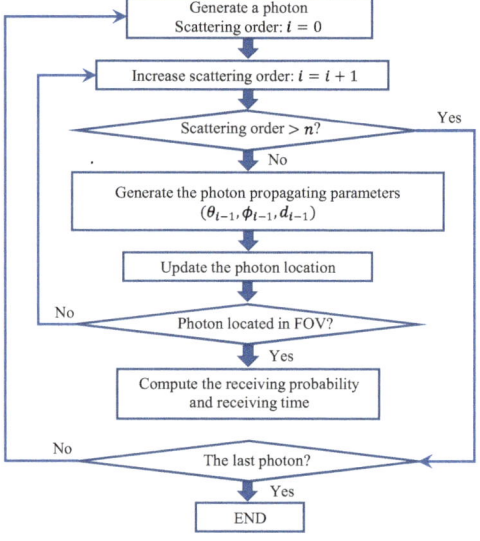

2.3 Numerical Results

In this section, we present some numerical results on channel path loss and CIR in typical transceiver geometries, where the transmitting beam axis and the receiving FOV axis are in coplanar case, the transmitting elevation angle and the receiving elevation angle are 45°, and communication distance is 100 m if not otherwise specified. The channel path loss is defined in unit of dB as $PL = 10 \times \lg 1/P_r$, where P_r is the receiving probability.

The path loss in dB under different communication distance is shown in Fig. 2.13. Here we present the results of both single-scattering model and multiple scattering model. From Fig. 2.13, we can observe that the channel path loss increases as the communication distance increases. Besides, we can also observe that when multiple scattering effect is considered, the channel path loss will decrease.

Then we present the CIR under typical system geometry with communication distance $r = 100$ m in Fig. 2.14. From Fig. 2.14 we can see that there exits a serious

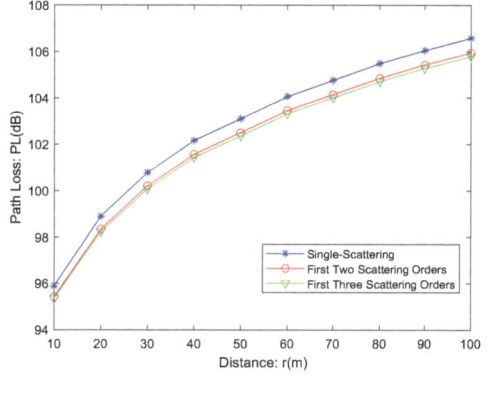

Fig. 2.13 Typical path losses under different communication distance

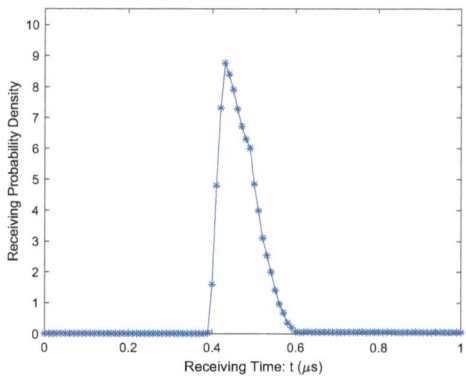

Fig. 2.14 Typical CIR of UV communication with communication distance $r = 100$ m

impulse spreading effect on the CIR. This impulse spreading effect will introduce serious inter-symbol interference on the UV communication systems, which is the major topic of the next chapter.

2.4 Summary and Future Directions

In this chapter, we introduced two types of channel models for NLOS UV communications, i.e., the analytical channel models and the stochastic channel models. For the analytical channel models, we introduced the single-scattering model, the second-order scattering model, and the third-order scattering model. Some approximated channel models were also discussed. For the stochastic channel models, we introduced the MCS model and the MCI model. Different types of sampling methods for MCI model were also provided. The analytical channel models enjoy fast computational speed but suffer from restricted application scenarios. Meanwhile, the stochastic channel models enjoy wide and flexible application scenarios but suffer from low computational speed.

There is still much room for improvement in existing channel models for NLOS UV communications. For example, more realistic light source distribution and reception models need to be considered [13]. At the same time, further research is needed for modeling in environments with obstacles and handling the reflection or absorption properties of obstacle surfaces. Some preliminary works can be found in [14, 15]. Moreover, stochastic channel models are often used in long-distance UV communication, thus the channel modeling also needs to consider the influence of atmospheric turbulence. Although there are some preliminary explorations in this direction [16–18], there is still a lot of room for improvement from practical turbulence channel models. Finally, the channel modeling method that combines stochastic channel models with analytical channel models is expected to combine the advantages of both. A pioneer work can be found in [19].

References

1. Luettgen, M.R., Reilly, D.M., Shapiro, J.H.: Non-line-of-sight single-scatter propagation model. J. Opt. Soc. A. A **8**(12), 1964–1972 (1991)
2. Elshimy, M.A., Hranilovic, S.: Non-line-of-sight single-scatter propagation model for non-coplanar geometries. J. Opt. Soc. A. A **28**(3), 420–428 (2011)
3. Zuo, Y., Xiao, H.F., Wu, J., et al.: A single-scatter path loss model for non-line-of-sight ultraviolet channels. Opt. Express **20**(9), 10359–10369 (2012)
4. Yin, H., Chang, S., Wwang, X., et al.: Analytical model of non-line-of-sight single-scatter propagation. J. Opt. Soc. A. A **27**(7), 1505–1509 (2010)
5. Wang, L., Xu, Z.Y., Sadler, B.M.: An approximate closed-form link loss model for non-line-of-sight ultraviolet communication in noncoplanar geometry. Opt. Lett. **36**(7), 1224–1226 (2011)

6. Wu, T.F., Ma, J.S., Yuan, R.Z., et al.: Single-scatter model for short-range ultraviolet communication in a narrow beam case. IEEE Photon. Technol. Lett. **31**(3), 265–268 (2019)
7. Yuan, R.Z., Ma, J.S., Su, P., et al.: An integral model of two-order and three-order scattering for non-line-of-sight ultraviolet communication in a narrow beam case. IEEE Commun. Lett. **20**(12), 2366–9 (2016)
8. Ding, H.P., Chen, G., Majumdar, A.K., et al.: Modeling of non-line-of-sight ultraviolet scattering channels for communication. IEEE J. Sel. Areas Commun. **27**(9), 1535–1544 (2009)
9. Drost, R.J., Moore, T.J., Sadler, B.M.: UV communications channel modeling incorporating multiple scattering interactions. J. Opt. Soc. A. A **28**(4), 686–695 (2011)
10. Ding, H.P., Xu, Z.Y., Sadler, B.M.: A path loss model for non-line-of-sight ultraviolet multiple scattering channels. EURASIP J. Wirel. Commun. Netw. **2010**, 63 (2010)
11. Yuan, R.Z., Ma, J.S., Su, P., et al.: Monte-Carlo integration models for multiple scattering based optical wireless communication. IEEE Tran. Commun. **68**(1), 334–348 (2020)
12. Tokdar, S.T., Kass, R.E.: Importance sampling: a review. Wiley Interdiscip. Rev.: Comput. Stat. **2**(1), 54–60 (2010)
13. Cao, T., Gao, X., Wu, T., et al.: Single-scatter path loss model of LED-based non-line-of-sight ultraviolet communications. Opt. Lett. **46**(16), 4013–4016 (2021)
14. Cao, T., Wu, T., Pan, C., et al.: Single-collision-induced path loss model of reflection-assisted non-line-of-sight ultraviolet communications. Opt. Express **30**(9), 15227–15237 (2022)
15. Cao, T., Gao, X., Wu, T., et al.: Reflection-assisted non-line-of-sight ultraviolet communications. J. Lightw. Technol. **40**(7), 1662–1665 (2021)
16. Ding, H.P., Chen, G., Majumdar, A.K., et al.: Turbulence modeling for non-line-of-sight ultraviolet scattering channels. Atmos. Propag. VIII **8038**, 195–202 (2011)
17. Wang, P., Xu, Z.Y.: Characteristics of ultraviolet scattering and turbulent channels. Opt. Lett. **38**(15), 2773–2775 (2013)
18. Shan, T., Ma, J.S., Wu, T.F., et al.: Single scattering turbulence model based on the division of effective scattering volume for ultraviolet communication. Chin. Opt. Lett. **18**(12), 120602 (2020)
19. Borah, D.K., Mareddy, V.R., Voelz, D.G.: Single and double scattering event analysis for ultraviolet communication channels. Opt. Express **29**(4), 5327–5342 (2021)

Chapter 3
Achievable Information Rates of Ultraviolet Communications

Abstract The strong scattering effects of UV signals in the atmosphere enable the NLOS ability of UV communications, but introduces large time dispersion of the channel impulse response (CIR), which inevitably results in serious inter-symbol interferences (ISIs). The effects of ISIs on communication performance can be quantified by using an analytical CIR function. In this chapter, through analyzing the impacts of coplanar and non-coplanar geometries on CIRs, a modified Gamma function (MGF) is presented for the precise analytical characterization of CIRs in the presence of multiple scattering. We first introduce an analytical CIR, named MGF, related to the system geometries in Sect. 3.1. Then Based on the MGF model, we quantify the ISI in Sect. 3.2. Then we derive the bit-error rate (BER) and achievable information rate (AIR) for both on-off keying (OOK) modulation and 2-pulse-position modulation (2-PPM) in Sect. 3.3. Section 3.4 presents some numerical results to verify the derived BER and AIR. Numerical results indicate that the MGF model provides higher accuracy in fitting both CIRs and BERs compared to the traditional Gamma function model. At last, we conclude this chapter in Sect. 3.5.

Keywords Channel impulse response · Inter-symbol interference · Achievable information rates

3.1 Analytical Channel Impulse Response

There are two methods to obtain the analytical channel impulse responses (CIR). First, the CIR with a multiple integral form can be obtained by using the analytical channel model, as introduced in Sect. 2.1. This type of analytical CIR has the advantage of high accuracy and short running time. However, it is challenging to find the integral limits of the integral form for the CIR because these integral limits are closely related to the system geometry. Second, the CIR with an closed-form function can be obtained by using the parameters fitting method. For example, a Gamma function (GF) was proposed to approximate the CIR of single-scattering models and

© The Author(s), under exclusive license to Springer Nature Singapore Pte Ltd. 2025   37
R. Yuan and Z. Wang, *Non-Line-of-Sight Ultraviolet Communications*,
SpringerBriefs in Computer Science, https://doi.org/10.1007/978-981-97-8543-8_3

a double-Gamma function was proposed to approximate the CIR of multiple scattering models. However, the influences of system geometries on the CIR were ignored in previous analytical CIR functions, which inevitably degrades the precision of the analytical CIR functions [1, 2]. Therefore, it is necessary to investigate the relation between the CIR and the system geometry and propose a more precise CIR function compared with existing analytical CIR functions for NLOS UV communications.

3.1.1 Relation Between CIRs and Common Volumes

According to the number of vertices of the common volume, NLOS UV system geometries can be classified into four cases,[1] i.e., the single vertex case, the double vertices case, the three vertices case and the four vertices case. The system geometries and the shapes of CIRs for these four cases are shown in 3.1. For the single vertex case, we can observe that the CIR in Fig. 3.1b is smooth and the only vertex A in Fig. 3.1a corresponds to the starting time instant A of the CIR in Fig. 3.1b. This is because the propagating path passing though point A is the shortest one among all the propagating paths. For the double vertices case, we can s that there is a sharp vertex B on the CIR in Fig. 3.1d, which corresponds to the vertex B of the common volume in Fig. 3.1c. This is because there is a sudden change on the common volume at vertex B; and the non-smooth spacial characteristics of the common volume will result in the non-derivation mathematical properties of the CIR function. Similarly, for the three vertices case, we can observe that there are two sharp vertices B and C on the CIR in Fig. 3.1f, which correspond to the vertices B and C of the common volume in Fig. 3.1e. However, we can see that the CIR at the vertex C is smoother than that at the vertex B. This is because the change rate of the volume at C is smaller than that at B. For the four vertices case, there is a vertex D in Fig. 3.1g, which corresponds to the ending time instant D of the CIR in Fig. 3.1h. This is because the propagating path passing though point D is the longest one among all the propagating paths.

From the above analysis on the relation between CIRs and common volumes, we can conclude that the shape of the CIR is significantly determined by the coordinates of the vertices. Therefore, in the following, we present the calculation of the coordinates of the vertices. The vertices are the intersections of the boundary lines of the transmitting beam and the receiving FOV, which can be expressed as

$$\begin{cases} \text{Line } T_1 : z_{T_1} = \tan\left(\vartheta_{T_1}\right) y, \quad z_{T_1} > 0; & (3.1a) \\ \text{Line } T_2 : z_{T_2} = \tan\left(\vartheta_{T_2}\right) y, \quad z_{T_2} > 0; & (3.1b) \\ \text{Line } R_1 : z_{R_1} = -\tan\left(\vartheta_{R_1}\right) y + r\tan(\vartheta_{R_1}), \quad z_{R_1} > 0; & (3.1c) \\ \text{Line } R_2 : z_{R_2} = -\tan\left(\vartheta_{R_2}\right) y + r\tan(\vartheta_{R_2}), \quad z_{R_2} > 0; & (3.1d) \end{cases}$$

---

[1] Here we only consider the coplanar cases. For non-coplanar cases, please refer to [2].

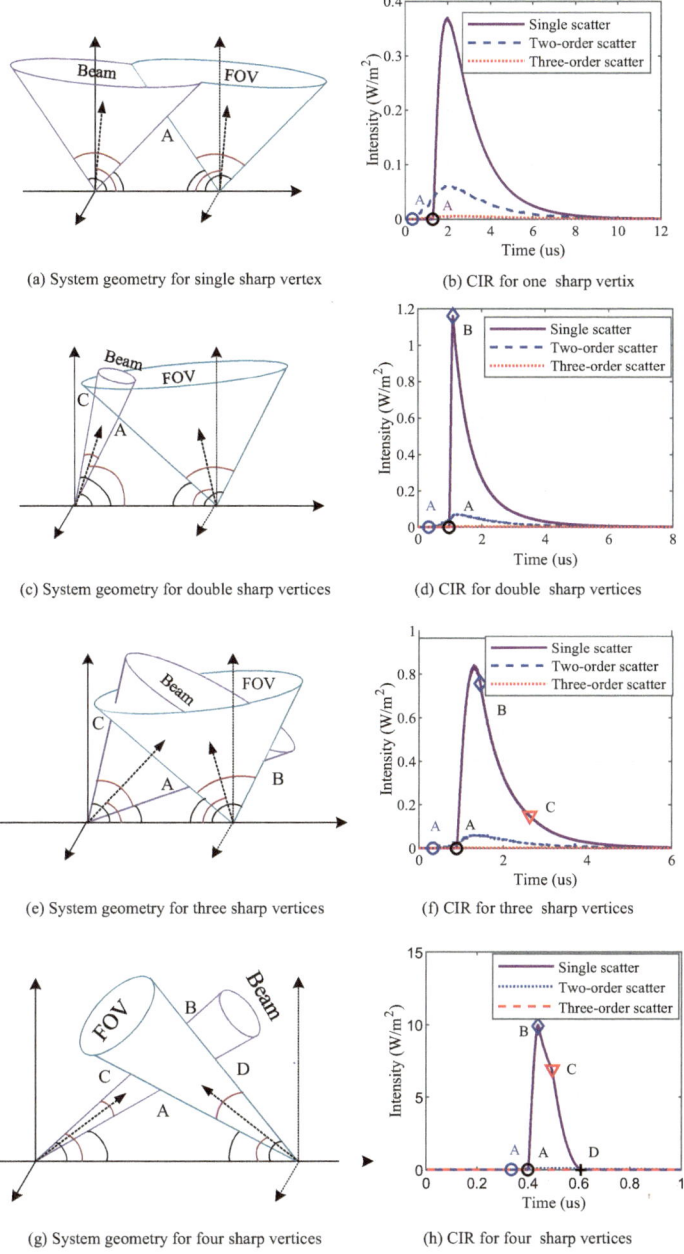

(a) System geometry for single sharp vertex

(b) CIR for one sharp vertix

(c) System geometry for double sharp vertices

(d) CIR for double sharp vertices

(e) System geometry for three sharp vertices

(f) CIR for three sharp vertices

(g) System geometry for four sharp vertices

(h) CIR for four sharp vertices

Fig. 3.1 Four cases of NLOS UV system geometries and the corresponding CIRs

where $\vartheta_{T_1} = \theta_T - \frac{\beta_T}{2}$ and $\vartheta_{T_2} = \theta_T + \frac{\beta_T}{2}$ are angles between left and right boundaries of the transmitting beam and the y-axis, respectively; $\vartheta_{R_1} = \theta_R - \frac{\beta_R}{2}$ and
$\vartheta_{R_2} = \theta_R + \frac{\beta_R}{2}$ are angles between left and right boundaries of the receiving FOV
and the y-axis, respectively. Then, the coordinates of vertices can be obtained by
solving the equations: $z_{T_1} - z_{R_1} = 0$; $z_{T_2} - z_{R_1} = 0$; $z_{T_1} - z_{R_2} = 0$; $z_{T_2} - z_{R_2} = 0$.
The solutions of these equations correspond to the coordinates of the vertices, A, B,
C, and D, respectively. The number of solutions indicates how many vertices exist
on the common volume.

We also presented the CIRs for the two-order and the three-order scattering in
Fig. 3.1. From Fig. 3.1, can observe that the intensity of three-order scattering is
much lower than those of the single scattering and the two-order scattering in typical
system geometries. Therefore, the single scattering and the two-order scattering are
only considered in this chapter. Besides, from Fig. 3.1, we observe that the CIR for
the two-order scattering is smooth and only a sharp vertex A is at the starting time
of the CIR.

3.1.2 Modified Gamma Function Model

We fully take the shapes of common volumes into consideration and classify the
CIRs into two types according to the shapes of CIRs, i.e., the single function fits the
smooth CIRs for the single and double vertices cases of the coplanar geometry and the
partial intersection case of the non-coplanar geometry; and the piecewise function fits
the non-smooth CIRs for the three and four vertices cases of the coplanar geometry
and the full intersection case of the non-coplanar geometry. For the first type, the
analytical CIR can be expressed as a single function

$$\hat{h}(t) = \frac{a}{(t - t_A)^{\frac{3}{2}}} \exp\left\{\frac{-[r - \varsigma(t - t_A)]^2}{b(t - t_A)}\right\}, \tag{3.2}$$

which is an MGF. Here b and ς are the fitting parameters, which determine the shape
of CIRs; a is the normalization constant; and $t_A = d_s/c$ is the starting time of CIRs,
where d_s is the shortest photon prorogating distance.

For the second type, the analytical CIR can be expressed as a piecewise function

$$\hat{h}(t) = \begin{cases} \hat{h}_1(t) : \frac{a_1}{(t - t_A)^{\frac{3}{2}}} \exp\left\{\frac{-[r - \varsigma(t - t_A)]^2}{b(t - t_A)}\right\}, & t_A \leq t < t_1 \quad \text{or} \quad t \geq t_2 \\ \hat{h}_2(t) : a_2(t - \tau)^2 + a_3, & t_1 \leq t < t_2, \end{cases} \tag{3.3}$$

which is formed by an MGF and a quadratic function. Here b, ς and τ are the fitting
parameters; a_1, a_2 and a_3 are normalization constants; $t_1 \triangleq \min\{t_B, t_C\}$ is defined as
the receiving time of the second vertex, and t_B and t_C correspond to the receiving

time of vertex B and vertex C, respectively. Similarly, $t_2 \triangleq \max\{t_B, t_C\}$ indicates the receiving time of the third vertex.

Based on the analysis of the CIR, here we present the CIR for the two-order scattering. Because the CIR for the two-order scattering is smooth, it can be fitted by using the single function (3.2), where the starting time is $t_A = r/c$. Then, the CIRs considering both the single scattering and the two-order scattering effects can be expressed as

$$\hat{h}(t) = \frac{H_{pl_s}}{H_{pl_s} + H_{pl_t}} \hat{h}_s(t) + \frac{H_{pl_t}}{H_{pl_s} + H_{pl_t}} \hat{h}_t(t), \tag{3.4}$$

where $\hat{h}_s(t)$ and $\hat{h}_t(t)$ are respectively the fitting CIRs for the single scattering and the two-order scattering effect; and H_{pl_s} and H_{pl_t} are respectively the path loss of the single scattering the two-order scattering.

3.1.3 Pseudocode of Curve Fitting of CIR

Then, we present the fitting process of the two types for the analytical CIRs (3.2) and (3.3). For the first type, the normalization constant a can be obtained by $\int_{t_A}^{\infty} \hat{h}(t)dt = H(\infty) - H(t_A) = 1$, where $H(t) \triangleq \int \hat{h}(t)dt$ can be obtained as

$$\begin{aligned} H(t) = fraca\sqrt{\pi} br Q & \left(\frac{\sqrt{2}r}{\sqrt{b(t-t_A)}} - \varsigma \sqrt{\frac{2(t-t_A)}{b}} \right) \sqrt{b(t-t_A)} \\ & + \frac{a\sqrt{\pi b}}{r} Q \left(\frac{\sqrt{2}r}{+} \varsigma \sqrt{\frac{2(t-t_A)}{b}} \right) e^{\frac{4r\varsigma}{b}} ; \end{aligned} \tag{3.5}$$

and $Q(x) = \int_x^{\infty} \frac{1}{\sqrt{2\pi}} e^{-0.5t^2} dt$ is the Q-function. Then we can obtain

$$a = \begin{cases} \frac{r}{\sqrt{\pi b}}, & \varsigma \geq 0; \\ \frac{re^{\frac{-4r\varsigma}{b}}}{\sqrt{\pi b}}, & \varsigma \leq 0. \end{cases} \tag{3.6}$$

Then the fitting parameters b and ς in (3.2) can be obtained by fitting the analytical CIR function to the numerical CIR results obtained by Monte Carlo simulations using fitting functions of numerical softwares, e.g., Matlab or Mathematica.

For the second type, we define $H_1(t) \triangleq H(t)$ and $H_2(t) \triangleq \int h_2(t)dt$, which can be obtained as

$$H_2(t) = \frac{1}{3}a_2(t-\tau)^3 + a_3(t-\tau). \tag{3.7}$$

Then, using the equations $\hat{h}_1(t_1) = \hat{h}_2(t_1)$, $\hat{h}_1(t_2) = \hat{h}_2(t_2)$, and $\int_{t_A}^{\infty} \hat{h}(t)\mathrm{d}t = 1$, we can obtain the normalization constant a_1, a_2, and a_3 in (3.3) as

$$a_1 = \begin{cases} \dfrac{1}{\bar{H}_1(t_1)-\bar{H}_1(t_2)+\frac{1}{3}A_1[(t_2-\tau)^3-(t_1-\tau)^3]+A_2(t_2-t_1)+\frac{r}{\sqrt{\pi b}}}, & \varsigma \geq 0, \\[4mm] \dfrac{1}{\bar{H}_1(t_1)-\bar{H}_1(t_2)+\frac{1}{3}A_1[(t_2-\tau)^3-(t_1-\tau)^3]+A_2(t_2-t_1)+\frac{r}{\sqrt{\pi b}}e^{\frac{-4r\varsigma}{b}}}, & \varsigma \leq 0 \end{cases} \tag{3.8}$$

$$a_2 = A_1 a_1,$$

$$a_3 = A_2 a_1,$$

where $\bar{H}_1(t) \triangleq H_1(t)/a_1$, $A_1 \triangleq [\bar{h}_1(t_1) - \bar{h}_1(t_2)]/[(t_1 - \tau)^2 - (t_2 - \tau)^2]$, $A_2 \triangleq \bar{h}_1(t) - A_1(t_1 - \tau)^2$, and $\bar{h}_1(t_1) \triangleq h_1(t)/a_1$. Similarly, the parameters b, ς and τ in (3.3) can be obtained by fitting the analytical CIR function to the numerical CIR results.

3.2 Inter-Symbol-Interference

We present both the OOK modulation and 2-PPM, where the transmitted bits are $\mathbf{x} = [x_0, x_1, \ldots, x_{L-1}]$, with $x_l \in \{0, 1\}$ referring to the lth ($l = 0, \ldots, L - 1$) bit; and L is the length of bits. The transmitted bits 0 and 1 are encoded by the absence and presence of an impulse for both OOK modulation and 2-PPM. Then the transmitted UV signal can be expressed as

$$x(t) = \sum_{l=0}^{L-1} x_l p(t - lT_b), \tag{3.9}$$

where T_b is the bit interval; $p(t)$ represents the transmitted shaping pulse. After passing through the NLOS UV communication channel, the received signal can be expressed as

$$y(t) = h(t) \otimes x(t) + \eta(t) = \sum_{l=0}^{L-1} E_b x_l g(t - lT_b) + \eta(t), \tag{3.10}$$

where $\otimes$ denotes the convolution operator; $h(t)$ is the CIR. E_b is the energy of the impulse; For simplicity, we move the $h(t)$ to the origin, i.e., $h(t) = \hat{h}(t + t_A)$; $g(t) = h(t) \otimes p(t)$ denotes the received waveform; $\eta(t)$ denotes the power of the thermal noise. Then the received signal at the lth bit time can be expressed as

$$y_l(t) = s_l(t) + i_l(t) + \eta(t), \quad t \in (lT_b, (l + 1)T_b]; \tag{3.11}$$

where $s_l(t) \triangleq E_b x_l g(t - lT_b)$ indicates the received information signal; $i_l(t) \triangleq \sum_{m=1}^{M} E_b x_{l-m} g(t - (l - m)T_b)$ indicates the ISI; $m \in \{1, 2, \dots, M\}$ denotes the index of ISIs. Here the maximum index $M = \lceil \frac{T_\tau}{T_b} \rceil - 1$; T_τ is the pulse width of the received signal and T_τ is defined by $T_\tau \triangleq \arg \min_{T_\tau} \int_0^{T_\tau} g(t)dt \geq 99.99\%$.

The detected number of photons n for the lth bit in the presence of ISI and thermal noise can be expressed as three parts, i.e., $n = n_s + n_i + n_\eta$, where n_s, n_i and n_η are the received number of photons due to the signal, ISI, and thermal noise, respectively. The distributions of n_s and n_i can be approximated as Poisson distributions [4–6]. Then the probability density for $n_{si} \triangleq n_s + n_i$ can be obtained as

$$p_{n_{si}}(k) = \frac{\bar{n}_{si}^k}{k!} e^{-\bar{n}_{si}}, \tag{3.12}$$

where k is the number of received photons; $\bar{n}_{si} = \bar{n}_s + \bar{n}_i$ is the average number of photons due to the signal and ISI. The ISI for UV communication can be modeled as a discrete-time finite-state channel [4]. Then $\bar{n}_{si}$ can be obtained as

$$\bar{n}_{si} = n_0 x_l q_0 + n_0 \sum_{m=1}^{M} x_{l-m} q_m = n_0 \sum_{m=0}^{M} x_{l-m} q_m, \tag{3.13}$$

where h is Plank constant; v is the frequency of the UV signal; $n_0 = \frac{E_b H_{pl}}{hv}$ is the number of received photons; H_{pl} is the path loss; q_m is probability of detecting an ISI photon due to the mth ISI, which can be expressed as

$$q_m = \int_0^{T_b} g(t + mT_b)dt; \tag{3.14}$$

and q_0 is the probability of detecting a photon due to the signal, which can be obtained by (3.14) for setting $m = 0$.

From (3.13), we see that the average number of photons is determined by q_m. Therefore, it is important to derive the analytical form of q_m for quantifying the channel ISIs. In the following, we derive q_m when the shaping pulse is the Delta function and the square wave, respectively.

3.2.1 Quantified the ISIs for Employing the Delta Function as the Shaping Pulse

We first give the analytical form of q_m when the transmitted shaping pulse is the Dirac delta function, i.e. $p(t) = \delta(t)$. Then, the received waveform is $g(t) = h(t)$; the probabilities q_m for (3.14) can be equivalently expressed as $q_m = \int_0^{T_b} h(t + mT_b)dt$.

When the number of vertices of the common volume is one or two for the coplanar geometry, or when the system geometry is non-coplanar and partial intersection, by substituting (3.2) into (3.14), we can obtain

$$q_m = \hat{H}(T_{m+1}) - \hat{H}(mT_b),$$ (3.15)

where $\hat{H}(t) = H(t + t_A)$; $T_{m+1} = (m + 1)T_b$ is the $(m + 1)$th bit time.

When the number of vertices of the common volume is three or four, or when the system geometry is non-coplanar and full intersection, by substituting (3.3) into (3.14), we can obtain

$$q_m = \begin{cases} \hat{H}_1(T_{m+1}) - \sum\limits_{j=0}^{m-1} q_j, & T_{m+1} \in (0, \hat{t}_1] \\ \hat{H}_1 \hat{H}_2(\hat{t}_1) + \hat{H}_2(T_{m+1}) - \sum\limits_{j=0}^{m-1} q_j, & T_{m+1} \in (\hat{t}_1, \hat{t}_2] \\ \hat{H}_1 \hat{H}_2(\hat{t}_1) - \hat{H}_1 \hat{H}_2(\hat{t}_2) + \hat{H}_1(T_{m+1}) - \sum\limits_{j=0}^{m-1} q_j, & T_{m+1} \in (\hat{t}_2, \infty) \end{cases}$$ (3.16)

where $\hat{H}_1(t) = \hat{H}(t)$; $\hat{H}_2(t) = H_2(t + t_A)$ is given in (3.7); $\hat{H}_1 \hat{H}_2(t) \triangleq \hat{H}_1(t) - \hat{H}_2(t)$; $\hat{t}_1 \triangleq \min\{t_B - t_A, t_C - t_A\}$ and $\hat{t}_2 \triangleq \max\{t_B - t_A, t_C - t_A\}$ are respectively the receiving time of the second vertex and the third vertex for moving the $h(t)$ to the origin.

3.2.2 Quantified the ISIs for Employing the Rectangular Function as the Shaping Pulse

Then we present the analytical form of q_m when using the rectangular function as the transmitted shaping pulse, i.e. $p(t) = u(t) - u(t - t_\mu)$, where $u(t)$ is the step function; t_μ is the pulse width with $t_\mu \le T_b$.

Similar to the Dirac delta function case, when the CIR curve is smooth, the received waveform can be expressed as $g(t) = h(t) \otimes p(t) = \hat{H}(t) - \hat{H}(t - t_\mu)$. Then the ISI can be obtained as

$$q_m = G(T_{m+1}) - G(mT_b) - G(T_{m+1} - t_\mu) + G(mT_b - t_\mu),$$ (3.17)

where $G(t) \triangleq \int_0^t \hat{H}(t) dt$. Similarly, when the CIR curve is non-smooth, the ISI can be obtained as

$$q_m = -\sum_{j=0}^{m-1} q_j +$$

$$
\begin{cases}
\Delta G_1(T_{m+1}, T_{m+1} - t_\mu), 0 < T_{m+1} \le \hat{t}_1 \\
\Delta G_1(\hat{t}_1, T_{m+1} - t_\mu) + \Delta G_2(T_{m+1}, \hat{t}_1) \\
\quad + [\hat{H}_1(\hat{t}_1) - G(\hat{t}_1)](T_{m+1} - \hat{t}_1), \hat{t}_1 < T_{m+1} \le \hat{t}_1 + t_\mu \\
\Delta G_1(\hat{t}_1 - t_\mu, \hat{t}_1) - \hat{H}_1(\hat{t}_1)t_\mu + \Delta G_2(T_{m+1}, T_{m+1} - t_\mu) + G_2(\hat{t}_1), \hat{t}_1 + t_\mu < T_{m+1} \le \hat{t}_2 \\
\Delta G_2(\hat{t}_2, T_{m+1} - t_\mu) + \Delta G_1(T_{m+1}, \hat{t}_2) \\
\quad + \hat{H}_1 \hat{H}_2(t_1)t_\mu - \hat{H}_1 \hat{H}_2(t_2)(T_{m+1} - \hat{t}_2), \hat{t}_2 < T_{m+1} \le \hat{t}_2 + t_\mu \\
\Delta G_2(t_1, t_2) + \Delta G_1(T_{m+1}, T_{m+1} - t_\mu) + (\hat{H}_1 \hat{H}_2(t_1) - \hat{H}_1 \hat{H}_2(t_2))t_\mu, T_{m+1} > \hat{t}_2 + t_\mu
\end{cases}
$$

$$(3.18)$$

where $G_1(t) = G(t)$, $G_2(t) \triangleq \int_0^t \hat{H}_2(x)\mathrm{d}x = \frac{a_2}{12}(t - \tau + t_A)^4 + \frac{a_3}{2}(t - \tau + t_A)^2$, $\Delta G_1(t_a, t_b) \triangleq G_1(t_a) - G_1(t_b)$, and $\Delta G_2(t_a, t_b) \triangleq G_2(t_a) - G_2(t_b)$.

Next we consider the detected number of photons due to thermal noise. The thermal noise is quantum mechanically described by a Gaussian thermal state; and the photon statistics of the Gaussian thermal state are described by a Geometric distribution [7–9]. Then, the probability density of n_η is given by

$$p_{n_\eta}(k) = \frac{\bar{n}_\eta^k}{(\bar{n}_\eta + 1)^{k+1}}, \qquad (3.19)$$

where $\bar{n}_\eta \triangleq \frac{1}{hv}\int_0^{T_b} \eta(t)\mathrm{d}t$ is the average number of photons raised by the random electron current due to the thermal noise; and here $\bar{n}_\eta$ can be measured in practical implementations.

We assume that the thermal noise is independent from the input signal strength. Then the probability density of n can be obtained as

$$p_n(k|\bar{n}_{si}) = \frac{\bar{n}_\eta^k}{(\bar{n}_\eta + 1)^{k+1}} e^{-\bar{n}_{si}} L_k\left(\frac{\bar{n}_{si}(\bar{n}_\eta + 1)}{\bar{n}_\eta}\right), \qquad (3.20)$$

where $L_k(x) \triangleq \sum_{i=0}^{k} \frac{x^i}{i!}$ is the summation of the first $k + 1$ term of the Taylor expansion for e^x.

3.3 Bit-Error Rates and Achievable Information Rates

Based on the probability density of received photons in (3.20), we will obtain the BER and the achievable information rate of NLOS UV communications in the presence of ISIs and thermal noises.

3.3.1 Bit-Error Rates

For the BER performance of OOK modulation, the transmitted information can be detected by comparing the number of photons measured by the receiver with the threshold K_{th}. For a given threshold K_{th}, the decoding BER can be expressed as

$$
P_e(K_{th}) = \frac{1}{2} - \frac{1}{2^{M+1}} \sum_{x_{l-1}^{l-M}} \sum_{k=0}^{K_{th}} \frac{\bar{n}_\eta^k}{(\bar{n}_\eta + 1)^{k+1}} \left[e^{-\bar{n}_{si}(0,x_{l-1}^{l-M})} L_k \left(\frac{\bar{n}_{si}(0, x_{l-1}^{l-M})(\bar{n}_\eta + 1)}{\bar{n}_\eta} \right) \right.
$$
$$
\left. - e^{-\bar{n}_{si}(1,x_{l-1}^{l-M})} L_k \left(\frac{\bar{n}_{si}(1, x_{l-1}^{l-M})(\bar{n}_\eta + 1)}{\bar{n}_\eta} \right) \right],
$$

(3.21)

where $p(x)$ indicates the priori probability of transmitted bits; x_{l-1}^{l-M} is defined as $\{x_{l-1}, \ldots, x_{l-M}\}$. The optimal threshold K_{th}^* can be obtained by solving the optimization problem

$$
K_{th}^* = \arg\min_{K_{th}} P_e(K_{th}).
$$
(3.22)

From (3.21), we can observe that as the threshold K_{th} increases, the second term of the decoding BER decreases, and the third term increases. The optimal threshold is the maximum K_{th} satisfying the decrement of the second term being greater than or equal to the increment of the third term when the threshold increases from K_{th} to $K_{th} + 1$. Therefore, the optimization problem (3.22) can be simplified as

$$
K_{th}^* = \max \quad K_{th}
$$
$$
\text{s.t.} \quad \sum_{x_{l-1}^{l-M}} e^{-\bar{n}_{si}(0,x_{l-1}^{l-M})} L_{K_{th}+1} \left(\frac{\bar{n}_{si}(0, x_{l-1}^{l-M})(\bar{n}_\eta + 1)}{\bar{n}_\eta} \right) \geq
$$
$$
\sum_{x_{l-1}^{l-M}} e^{-\bar{n}_{si}(1,x_{l-1}^{l-M})} L_{K_{th}+1} \left(\frac{\bar{n}_{si}(1, x_{l-1}^{l-M})(\bar{n}_\eta + 1)}{\bar{n}_\eta} \right).
$$
(3.23)

The optimal threshold K_{th}^* can be obtained by using an exhaustive searching method.

In short-range NLOS UV communications, the ISI between adjacent bits is usually the main ISI that affecting the system performance. If only considering the ISI between adjacent bits, the BER (3.21) can be rewritten as

$$
P_e(K_{th}) = \frac{1}{2} - \frac{1}{4} p_n(k \le K_{th}|\bar{n}_{si} = 0) - \frac{1}{4} p_n(k \le K_{th}|\bar{n}_{si} = n_0 q_1)
$$
$$
+ \frac{1}{4} p_n(k \le K_{th}|\bar{n}_{si} = n_0 q_0) + \frac{1}{4} p_n(k \le K_{th}|\bar{n}_{si} = n_0(q_0 + q_1)).
$$
(3.24)

Noting that when $K_{th} = K_{th}^*$, we have $p_n(k \le K_{th}^*|\bar{n}_{si} = 0) \to 1$ and $p_n(k \le K_{th}^*|\bar{n}_{si} = n_0(q_0 + q_1)) \to 0$. We can obtain a lower bound of P_e by replacing

$p_n(k \leq K_{\text{th}}^* | \bar{n}_{si} = 0)$ and $p_n(k \leq K_{\text{th}}^* | \bar{n}_{si} = n_0(q_0 + q_1))$ with 1 and 0, respectively, i.e., we have

$$
P_e^{\text{lower}}(K_{\text{th}}^*) = \frac{1}{4} - \sum_{k=0}^{K_{\text{th}}^*} \frac{\bar{n}_\eta^k}{4(\bar{n}_\eta + 1)^{k+1}} e^{-n_0 q_1} L_k \left(\frac{n_0 q_1 (\bar{n}_\eta + 1)}{\bar{n}_\eta} \right)
$$
$$
+ \sum_{k=0}^{K_{\text{th}}^*} \frac{\bar{n}_\eta^k}{4(\bar{n}_\eta + 1)^{k+1}} e^{-n_0 q_0} L_k \left(\frac{n_0 q_0 (\bar{n}_\eta + 1)}{\bar{n}_\eta} \right).
$$

$$(3.25)$$

We note that the average number of received photons increases as the thermal noise $\bar{n}_\eta$ increases for both transmitted bits 0 and 1, which indicates that the optimal threshold K_{th}^* increases with the thermal noise $\bar{n}_\eta$. Therefore, the optimal threshold when $\bar{n}_\eta = 0$ can be regarded as a lower bound of the optimal threshold. The optimal threshold when $\bar{n}_\eta = 0$ can be obtained by solving the following optimization problem:

$$
K_{\text{th}}^{\text{lower}} = \max \ K_{\text{th}}
$$
$$
\text{s.t.} \ (n_0 q_1)^{K_{\text{th}}+1} e^{-n_0 q_1} - (n_0 q_0)^{K_{\text{th}}+1} e^{-n_0 q_0} \geq 0.
$$

$$(3.26)$$

The solution of (3.26) can be obtained as $K_{\text{th}}^{\text{lower}} = \left\lfloor \frac{n_0 q_0 - n_0 q_1}{\ln q_0 - \ln q_1} \right\rfloor$, where $\lfloor \cdot \rfloor$ represents the greatest integer floor function.

For the BER performance of 2-PPM, the information bit can be detected by comparing the measured number of photons at the first and second halves in each information bit time. The detection rule can be expressed as $\hat{x} = 0$ for $n_l \geq n_{l+1}$ and $\hat{x} = 1$ for $n_l < n_{l+1}$, where n_l and n_{l+1} are the number of photons measured at the first and the second halves of an information bit time, respectively. Then the decoding BER for 2-PPM can be expressed as

$$
P_e
$$
$$
= 1 - \sum_{x_i} \sum_{k=0}^{\infty} \frac{p_n \left(k | \bar{n}_{si,l}(x_{l+1} = 0, x_l = 1, \mathbf{x}_i) \right)}{2 \lfloor (M-1)/2 \rfloor + 2} p_n \left(\kappa \leq k | \bar{n}_{si,l+1}(x_{l+1} = 0, x_l = 1, \mathbf{x}_i) \right)
$$
$$
- \sum_{x_i} \sum_{k=0}^{\infty} \frac{p_n \left(k | \bar{n}_{si,l+1}(x_{l+1} = 1, x_l = 0, \mathbf{x}_i) \right)}{2 \lfloor (M-1)/2 \rfloor + 2} p_n \left(\kappa \leq k | \bar{n}_{si,l}(x_{l+1} = 1, x_l = 0, \mathbf{x}_i) \right),
$$

$$(3.27)$$

where $\mathbf{x}_i = [x_{l-1}, \ldots, x_{l-M}]$ indicates the ISI transmitted bits; $\bar{n}_{si,l}$ and $\bar{n}_{si,l+1}$ are the average number of photons due to the signal and ISI, respectively. If we only consider the ISI between adjacent information bits, the decoding BER (3.27) can be rewritten as

$$P_e = 1 - \sum_{k=0}^{\infty} \frac{p_n(k|n_0(q_0 + q_2))}{2} p_n \left(\kappa \leq k|n_0 q_1 \right) - \sum_{k=0}^{\infty} \frac{p_n(k|n_0 q_0)}{4} p_n \left(\kappa \leq k|n_0 q_2 \right)$$

$$- \sum_{k=0}^{\infty} \frac{p_n(k|n_0(q_0 + q_1))}{4} p_n \left(\kappa \leq k|n_0(q_1 + q_2) \right).$$

$$(3.28)$$

Noting that when $n_0 q_0 \gg n_0 q_2$, we have that both $\sum_{k=0}^{\infty} p_n(k|n_0(q_0 + q_2))$ $p_n \left(\kappa \leq k|n_0 q_1 \right)$ and $\sum_{k=0}^{\infty} p_n(k|n_0 q_0) p_n \left(\kappa \leq k|n_0 q_2 \right)$ approach 1. Then by substituting (3.20) into (3.28), we can obtain a lower bound of P_e^{lower} as

$$P_e^{\text{lower}} = \frac{1}{2} - \sum_{k=0}^{\infty} \frac{\bar{n}_\eta^k}{2(\bar{n}_\eta + 1)^{k+1}} e^{-n_0(q_0 + q_2)} L_k \left(\frac{n_0(q_0 + q_2)(\bar{n}_\eta + 1)}{\bar{n}_\eta} \right)$$

$$(3.29)$$

$$\times \sum_{\kappa=0}^{k} \frac{\bar{n}_\eta^\kappa}{2(\bar{n}_\eta + 1)^{\kappa+1}} e^{-n_0 q_1} L_\kappa \left(\frac{n_0 q_1(\bar{n}_\eta + 1)}{\bar{n}_\eta} \right).$$

3.3.2 Achievable Information Rates

Define $\mathcal{X}$ and $\hat{\mathcal{X}}$ as the input and output random processes of the channel, respectively. Additionally, we define $\mathbf{X}^K \triangleq [\mathbf{x}_0, \mathbf{x}_1, \ldots, \mathbf{x}_{K-1}]$ as the collection of K blocks of transmitted vectors length J [4] and define $\hat{\mathbf{X}}^K$ as the received collection of the transmitted blocks. Then we have $J = 1$ for OOK modulation and $J = 2$ for 2-PPM.

The achievable information rate describes the average amount of information transmitted per unit time, which can be defined as

$$R_I = R_t \lim_{K \to \infty} \frac{1}{JK} I(X^K, \hat{X}^K) = \frac{R_t}{J} [H(\hat{\mathcal{X}}) - H(\hat{\mathcal{X}}|\mathcal{X})], \qquad (3.30)$$

where R_I is the achievable information rate; $R_t = \frac{1}{T_b}$ is the transmission bit rate; $H(\hat{\mathcal{X}})$ and $H(\hat{\mathcal{X}}|\mathcal{X})$ are the entropy rates in units of bitblock [4].

When $\mathcal{X}$ and $\hat{\mathcal{X}}$ are stationary ergodic random processes, the entropies $H(\hat{\mathcal{X}})$ and $H(\hat{\mathcal{X}}|\mathcal{X})$ can be respectively estimated as $- \lim_{K \to \infty} \frac{1}{K} \log_2 \left[p(\hat{\mathbf{X}}^K) \right]$ and $- \lim_{K \to \infty} \frac{1}{K}$ $\log_2 \left[p(\hat{\mathbf{X}}^K|\mathbf{X}^K) \right]$ according to the Shannon-McMillan-Breiman theorem [10]. Therefore, the channel capacity can be estimated as

$$I(\mathcal{X}; \hat{\mathcal{X}}) \approx \frac{1}{J} \left\{ - \lim_{K \to \infty} \frac{1}{K} \log_2 \left[p(\hat{\mathbf{X}}^K) \right] + \lim_{K \to \infty} \frac{1}{K} \log_2 \left[p(\hat{\mathbf{X}}^K|\mathbf{X}^K) \right] \right\},$$

$$(3.31)$$

where the joint probability $p(\hat{\mathbf{X}}^K)$ and the conditional probability $p(\hat{\mathbf{X}}^K|\mathbf{X}^K)$ can be estimated numerically using the Monte-Carlo method.

For the achievable information rate, as the bit interval is shortened, more bits can be transmitted per unit time, and the transmitted rate R_t will be increased. However, when shortening the bit interval T_b, more serious ISIs will be introduced, which will degrade the BER performance and reduce the channel capacity. Therefore, there exists an optimal transmission bit rate that maximizing the achievable information rate. The optimization problem for obtaining the optimal achievable rate can be expressed as

$$\max_{R_t} R_t I(\mathcal{X}; \hat{\mathcal{X}}), \tag{3.32}$$

which can be solved by using numerical methods.

When ISIs are negligible, we can regard the channel as a memoryless channel. The channel capacity of the memoryless BSC can be obtained as

$$I(\mathcal{X}; \hat{\mathcal{X}}) = 1 + \log_2(1 - P_e^{\text{BSC}}), \tag{3.33}$$

where P_e^{BSC} is the BER of the BSC. When the ISI is small, we can approximate the channel capacity by replacing P_e^{BSC} in (3.33) with the BER in (3.24), where the ISI with adjacent bits are considered. Therefore, the achievable information rate can be approximated as

$$R_I \approx R_t[1 + \log_2(1 - P_e)], \tag{3.34}$$

which can greatly reduce the computation complexity compared with the rate (3.31) by Monte-Carlo methods.

3.4 Numerical Results

In this section, we evaluate the CIRs by using the MGF model and compare the CIRs with those obtained by the widely used GF model [11]. Unless otherwise specified, the system parameters are based on a typical experiment for NLOS UV communications with $\{\beta_T, \beta_R\} = \{17°, 30°\}$ [12]. For the atmospheric condition, we adopt the atmospheric scattering coefficient k_s is $0.55 \times 10^{-3} \text{m}^{-1}$. The atmospheric absorption coefficient k_a is $0.802 \times 10^{-3} \text{m}^{-1}$. The atmospheric extinction coefficient $k_e = k_s + k_a$ is $1.352 \times 10^{-3} \text{m}^{-1}$. The receiving area of the detector is $A_r = 1.77 \times 10^{-4}$; and we assume that the average photon n_0 of the received signal for a bit period is 100. Then the receiving energy per square meter can be estimated as $n_0 h \nu / A_r = 4.41 \times 10^{-13} \ J/m^2$, which is far more smaller than the threshold limit value of $60 \ J/m^2$ established by the American Conference of Governmental Industrial Hygienists [13].

The CIRs obtained by the MGF model, the GF model, and the single-scattering model under different system geometries are shown in Fig. 3.2. Figure 3.2a show the fitting results of the single vertex case; Fig. 3.2b show the fitting results of the double vertices case; Fig. 3.2c show the fitting results of the three vertices case; and

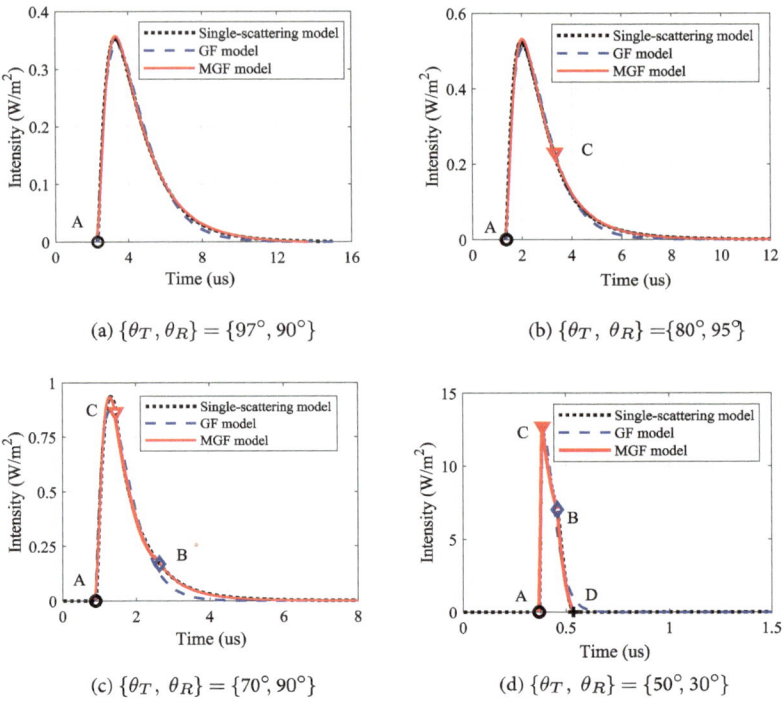

(a) $\{\theta_T, \theta_R\} = \{97°, 90°\}$

(b) $\{\theta_T, \theta_R\} = \{80°, 95°\}$

(c) $\{\theta_T, \theta_R\} = \{70°, 90°\}$

(d) $\{\theta_T, \theta_R\} = \{50°, 30°\}$

Fig. 3.2 The CIRs obtained by the MGF model, the GF model, and the single-scattering model [2]

Fig. 3.2d show the fitting results of the four vertices case. In the single vertex case and the second vertices case, we can see that the CIRs are smooth and there exists obvious fitting errors at the tail of the CIRs obtained by the GF model. In the three vertices case and the four vertices case, we can see that the CIRs are no longer smooth and a clear bump appears on the CIRs. Considering the influences of the bump on the CIRs, we use a piecewise function (3.3) for the MGF model to fit the non-smooth CIRs in Sect. 3.1.2. However, the GF model ignores the influence of system geometries, which will inevitably introduce serious fitting errors and further result in large estimation errors of the ISIs and BER.

In Fig. 3.3, we present the BER performance versus the transmission bit rate for OOK modulation and 2-PPM. We can see that although the difference in the CIR between the MGF model and the GF model in Fig. 3.2 looks relative small, there exists serious estimation errors for the BER obtained by the GF model. This is because the ISIs are determined by the tail of the CIRs; and from Fig. 3.2 we can observe that the major differences between the MGF model and the GF model appear mainly at the tail of the CIRs. We also present the lower bound of the BER obtained by (3.25) for OOK modulation and (3.29) for 2-PPM. We can see that the lower bound will approach the BER obtained by the MGF model as the transmission bit rate increases. Therefore, the lower bound can be used to roughly estimate the BER performance.

Fig. 3.3 Comparison of the BER performance between the MGF model and the GF model under the typical system geometries with $\{\theta_T, \theta_R\} = \{50°, 30°\}$ [2]

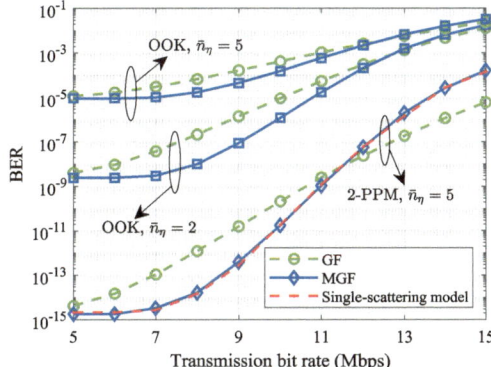

Besides, we can see that the BER performances for 2-PPM are better than that for OOK modulation, which prove that 2-PPM can achieve a better anti-interference ability than OOK modulation.

At last, we present the simulation results of the achievable information rate versus the transmission bit rate under different thermal noise levels, shown in Fig. 3.4. We can observe that under different noise levels, the achievable the information rate will first increase then decrease as the transmission bit rate increases. This is because the influence of ISIs on the achievable information rate is slight when the transmission bit rate is small. In this case, the achievable information rate is proportional to the transmission bit rate. As the transmission bit rate increases, the influence of ISIs becomes pronounced. Therefore, there exists an optimal transmission bit rate corresponding to the maximum achievable information rate. Besides, by comparing the achievable information rate between OOK modulation and 2-PPM given in Fig. 3.4, we can find that OOK modulation can achieve a higher maximum achievable information rate. This is because the encoding efficiency for 2-PPM is half of that for OOK modulation.

Fig. 3.4 The achievable information rate with different transmission bit rates under two thermal noise levels: the parameter of the system geometry is $\{\theta_T, \theta_R\} = \{50°, 30°\}$ [2]

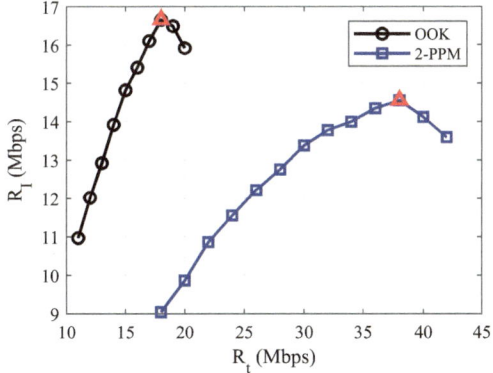

3.5 Summary and Future Directions

In this chapter, an MGF model was presented for analytically estimating the CIR of multiple scattering effects under coplanar system geometries. Based on the MGF model, both the ISIs and the probability density of the received photons were derived by considering the influence of channel ISIs, Poisson noise, and thermal noise. Then, the BER and the achievable information rate for both OOK modulation and 2-PPM were presented under a photon-counting receiver.

Recently, literatures [14, 15] have studied the channel characteristics of UV communication in mobile scenarios. It was found that the time dispersion of the CIR is widened in mobile scenarios. Besides, the time variation of the system geometry can also affect the shape of the CIR. Therefore, in the future, it is meaningful to extend the analysis to the mobile scenarios by considering the effects of motion of either source, destination or both.

References

1. Yuan, R.Z., Ma, J.S., Su, P., et al.: Monte-Carlo integration models for multiple scattering based optical wireless communication. IEEE Tran. Commun. **68**(1), 334–348 (2020)
2. Wang, Z., Yuan, R., Peng, M.: Inter-symbol interferences deteriorated ultraviolet communications using photon-counting receivers. IEEE Trans. Wirel. Commun. (2023)
3. Xu, F., Khalighi, A., Ali, A., et al.: Channel coding and time-diversity for optical wireless links. IEEE Commun. Lett. **17**(2), 872–887 (2009)
4. El-Shimy, M., Hranilovic, S.: Binary-input non-line-of-sight solar-blind UV channels: modeling, capacity and coding. J. Opt. Commun. Netw. **4**(12), 1008–1017 (2012)
5. El-Shimy, M., Hranilovic, S.: On the use of photon arrival-times for non-line-of-sight solar-blind UV channels. IEEE Commun. Lett. **18**(6), 913–916 (2014)
6. Zou, D., Gong, C., Wang, K., et al.: Characterization on practical photon counting receiver in optical scattering communication. IEEE Trans. Commun. **67**(3), 2203–2217 (2018)
7. Glauber, R.: The quantum theory of optical coherence. Phys. Rev. **130**(6), 2529 (1963)
8. Glauber, R.: Coherent and incoherent states of the radiation field. Phys. Rev. **131**(6), 2766 (1963)
9. Cariolaro, G.: Quantum Communications. Springer, Berlin (2015)
10. Algoet, P., Cover, T.: A sandwich proof of the Shannon-Mcmillan-Breiman theorem. Ann. Probab. **16**(2), 2203–2217 (1988)
11. Ding, H., Chen, G., Majumdar, A., et al.: Modeling of non-line-of-sight ultraviolet scattering channels for communication. IEEE J. Sel. Areas Commun. **27**(9), 1535–1544 (2009)
12. Chen, G., Abou-Galala, F., Xu, Z., et al.: Experimental evaluation of led-based solar blind NLOS communication links. Opt. Express **16**(19), 15059–15068 (2008)
13. Vavoulas, A., Sandalidis, H., Chatzidiamantis, N., et al.: A survey on ultraviolet c-band (UV-C) communications. IEEE Commun. Surv. Tuts. **21**(3), 2111–2133 (2019)
14. Song, P., Liu, C., Zhao, T., et al.: Research on pulse response characteristics of wireless ultraviolet communication in mobile scene. Opt. Express **27**(8), 10670–10683 (2019)
15. Song, P., Ji, H., Geng, X., et al.: On-channel characteristics of wireless ultraviolet communication with mobile terminals. Opt. Lett. **47**(4), 929–932 (2022)

Chapter 4
Full-Duplex Ultraviolet Communications

Abstract The full-duplex optical communications can be achieved simply by separating the transmitting link and the receiving link in the space. However, it is challenging to achieve the non-line-of-sight (NLOS) full-duplex ultraviolet (UV) communication due to the serious self-interference caused by the strong multiple scattering effects of UV signals. In this chapter, we first describe the self-interference channel by using an analytical channel impulse response (CIR) function and quantify the self-interference based on the CIR in Sect. 4.1. Then, based on the quantified self-interference, we present the bit-error rate (BER) and the corresponding achievable information rate (AIR) for on-off keying modulation and 4-pulse-position modulation in Sect. 4.2. Furthermore, We introduce a self-interference cancelling (SIC) method to mitigate the impacts of self-interferences in Sect. 4.3. In Sect. 4.4, we present some numerical results to explore the BER and AIR performance of full-duplex UV communication. Simulation results show that the SIC method can significantly improve the error rate and AIR performances. At last, we conclude this chapter in Sect. 4.5.

Keywords Full-duplex · Self-interference · Self-interference cancelling

4.1 Quantization of Self-Interference

4.1.1 Self-Interference Channel

The system geometry of the NLOS full-duplex UV communication is shown in Fig. 4.1. The communication links between two users are separated in the space. Without loss of generality, we assume that the two users adopt the same transmitting and receiving geometries and the elevation angles and the azimuth angles are symmetrical along the y-o-z plane. The single scattering links between the transmitters (Tx) and the receivers (Rx) are used to exchange the information bits. Due to the strong scattering effects of UV signals, the receiving link of one user will surfer from serious self-interference from the transmitting link of itself. Because the

© The Author(s), under exclusive license to Springer Nature Singapore Pte Ltd. 2025   53
R. Yuan and Z. Wang, *Non-Line-of-Sight Ultraviolet Communications*,
SpringerBriefs in Computer Science, https://doi.org/10.1007/978-981-97-8543-8_4

Fig. 4.1 The system
geometry of the single
scattering model for the
information channel

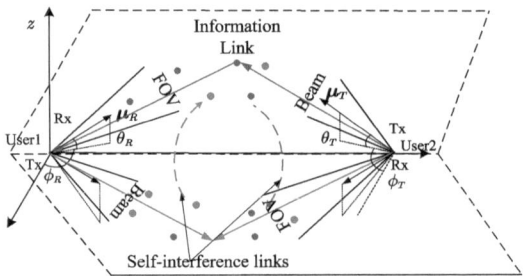

self-interference mainly comes from the two-order scattering effects, we can use the
two-order scattering model to characterize the self-interference channel.

We have presented the single scattering model for the information channel in
Chap. 2. Then we introduce the two-scattering model for the self-interference
channel. The geometrical setting for the two-order scattering process is shown in
Fig. 4.2, where the separating distance between the Tx and the Rx of the same
user is denoted by r_d; the direction cosines of the transmitting beam axis and the
receiving FOV axis are denoted by $\boldsymbol{\mu}_T$ and $\boldsymbol{\mu}_R$, respectively. The angle between
$\boldsymbol{\mu}_T$ and $\boldsymbol{\mu}_R$ is denoted by 2α; the elevation of the $\boldsymbol{\mu}_T$-o-$\boldsymbol{\mu}_R$ plane is denoted by
ϑ. The direction cosines are defined as $\boldsymbol{\mu}_T \triangleq [\cos\theta_T \cos\phi_T, \cos\theta_T \sin\phi_T, \sin\theta_T]^T$
and $\boldsymbol{\mu}_R \triangleq [\cos\theta_R \cos\phi_R, -\cos\theta_R \sin\phi_R, \sin\theta_R]^T$, where $\{\theta_T, \phi_T, \theta_R, \phi_R\}$ can be
obtained as

$$\begin{cases} \theta_T = \theta_R = \arcsin(\cos\alpha\sin\vartheta) \\ \phi_T = -\phi_R = \arcsin\left[\dfrac{\cos\alpha\cos\vartheta}{\sqrt{1-(\cos\alpha\sin\vartheta)^2}}\right]. \end{cases} \qquad (4.1)$$

Assume that the emitting process is the zero-order scattering, and the propagating
distance, scattering zenith angle and scattering azimuth angle of ith scattering are
respectively denoted by d_i, θ_i and ϕ_i, where $i = 0, 1, 2$. Following the derivation of

Fig. 4.2 The system
geometry of the two-order
scattering model for the
self-interference channel

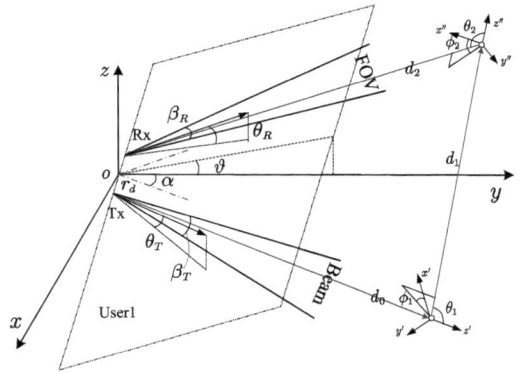

the Monte-Carlo integration (MCI) model in [1], we can obtain the receiving power of the two-order scattering as

$$P_{r,2} = P_t \int_V p_d \frac{k_s}{k_e} \left[\prod_{i=0}^{2} f_D(d_i) f_{\Theta_i}(\theta_i) f_{\Phi_i}(\phi_i) \right] \mathrm{d}d_0 \mathrm{d}\theta_0 \mathrm{d}\phi_0 \mathrm{d}d_1 \mathrm{d}\theta_1 \mathrm{d}\phi_1, \quad (4.2)$$

where p_d, $f_D(d_i)$, $f_{\Theta_i}(\theta_i)$, $f_{\Phi_i}(\phi_i)$, and V are defined in [1]. Accordingly, the channel path loss can be expressed as $PL_2 = -10 \log_{10} \frac{P_{r,2}}{P_t}$.

The CIR $h_2(t)$ for the two-order scattering can be obtained by using the MCI model or the Monte-Carlo simulation (MCS) model [1]. However, the CIR obtained by the MCI model or the MCS model is not an analytical form, which cannot be used for obtaining an analytical self-interference for UV communications. To facilitate the analysis of the self-interference, here we present a power function model to fit the CIR $h_2(t)$ obtained by the MCI model as

$$h_2(t) = a(t - b)^p, \quad (4.3)$$

where a, b, and p can be obtained by fitting the power function model to the numerical results of the MCI model though minimizing the normalized mean-square error (NMSE) criterion.

The distance of the information link is much longer than that of the self-interference link. It will cause that the arriving time of the information signal and the self-interference is different. Therefore, we have to consider the impact of the asynchronous arriving times for the information bits and the interference bits. Besides we can observe that the majority of the self-interference power are located closely to the origin. Therefore, for simplicity, we only consider the influence of the self-interference on two adjacent information bits in the following analysis.

4.1.2 Self-Interference Quantization

We introduce both OOK modulation and 4-PPM, where the information symbols are modulated into the transmission bits $x = [x_{s,0}, x_{s,1}, \ldots, x_{s,L-1}]$ with $x_{s,l} \in \{0, 1\}$ referring to the lth $(l = 0, \ldots, L - 1)$ bit; and L is the length of transmitted bits. The transmitted bits 0 and 1 are encoded by the absence and presence of an impulse. The transmitted UV signal can be expressed as

$$x_s(t) = \sum_{l=0}^{L-1} E_b x_{s,l} p(t - lT_b), \quad (4.4)$$

where $E_b = P_t T_b$ is the energy of the transmitted bit; T_b is the transmitted bit interval; $p(t)$ represents the pulse shape. After passing through the NLOS UV communication

channel, the received signals combined with both the information signal and the
self-interference can be expressed as

$$y(t) = h_s(t) \otimes x_s(t) + h_i(t) \otimes x_i(t - T_\tau) + \eta_b(t), \tag{4.5}$$

where $\otimes$ denotes the convolution operator; $x_i(t) = \sum_{l=0}^{L-1} E_b x_{i,l} p(t - lT_b)$ is the
transmitted self-interference signal; $h_s(t) = h_1(t) \otimes p(t)$ and $h_i(t) = h_2(t) \otimes p(t)$
are the received waveform for the information and the self-interference, respectively;
$\eta_b(t)$ denotes the background noise; T_τ is the asynchronous time between the information bit and the self-interference bit; and we define the asynchronous rate as $\tau \triangleq$
T_τ / T_b. We present the case of $T_b \geq T_m$, such that the impact of inter-symbol interferences on the performance of the UV communication can be ignored [2–8], where T_m is
the maximum duration of $h_s(t)$ and is defined as $\triangleq \arg\min_{T_m} \int_0^{T_m} h_s(t)dt \geq 99.99\%$.
Then substituting (4.4) into (4.5), it can obtain

$$y(t) = \sum_{l=0}^{L-1} E_b x_{s,l} h_s(t - lT_b) + \sum_{l=0}^{L-1} E_b x_{i,l} h_i(t - lT_b - T_\tau) + \eta_b(t), \tag{4.6}$$

where $x_{i,l}$ is the lth self-interference bit. Then the received signal at the lth bit time
can be expressed as

$$y_l(t) = s_l(t) + i_l(t) + i_{l-1}(t) + \eta_b(t), \quad t \in (lT_b, (l+1)T_b]; \tag{4.7}$$

where $s_l(t)$ indicates the received information signal; $i_l(t)$ and $i_{l-1}(t)$ are the synchronous self-interference signal and the asynchronous self-interference signal,
respectively.

Then the average received photons given the information bit $x_{s,l}$ and the self-
interference bits $\{x_{i,l}, x_{i,l-1}\}$ can be obtained as

$$\lambda(x_{s,l}, x_{i,l}, x_{i,l-1}) = \frac{1}{h\nu} \int_{lT_b}^{(l+1)T_b} y_l(t)dt = x_{s,l}\lambda_s + x_{i,l}\lambda_{i,1} + x_{i,l-1}\lambda_{i,2} + \lambda_b, \tag{4.8}$$

where h is Plank constant; ν is the frequency of the UV signal; λ_b is the average
number of photons for background noises; λ_s is the average number of photons
due to the information; $\lambda_{i,1}$ and $\lambda_{i,2}$ are the average number of photons due to the
synchronous self-interference and the asynchronous self-interference, respectively.

When the pulse shape is the delta function, the received waveform for the
information channel is $h_s(t) = h_1(t)$; and λ_s can be obtained as

$$\lambda_s = \frac{E_b}{h\nu} \int_{lT_b}^{(l+1)T_b} h_1(t - lT_b)dt = \frac{P_{r,1}T_b}{h\nu}. \tag{4.9}$$

The received waveform for the self-interference channel is $h_i(t) = h_2(t)$; $\lambda_{i,1}$ and
$\lambda_{i,2}$ can be obtained as

$$\lambda_{i,1} = \frac{E_b}{h\nu} \int_0^{T_b - T_\tau} h_2(t)\mathrm{d}t = \frac{E_b}{h\nu} \left[\frac{a}{p+1}(T_b - T_\tau - b)^{p+1} - \frac{a}{p+1}(-b)^{p+1} \right],$$

$$\lambda_{i,2} = \frac{E_b}{h\nu} \int_{T_b - T_\tau}^{T_b} h_2(t)\mathrm{d}t = \frac{E_b}{h\nu} \left[\frac{a}{p+1}(T_b - b)^{p+1} - \frac{a}{p+1}(T_b - T_\tau - b)^{p+1} \right].$$

$$(4.10)$$

When the pulse shape is the rectangular function, the received waveform for the information channel is $h_s(t) = h_1(t) \otimes p(t) = \int_{t - t_\mu}^{t} h_1(t)\mathrm{d}t$, where $p(t) = u(t) - t(t - t_\mu)$; $u(t)$ is the step function; t_μ is the pulse width with $t_\mu < T_b - T_m$. Then λ_s can be obtained as

$$\lambda_s = \frac{E_b}{h\nu} \int_{lT_b}^{(l+1)T_b} h_s(t - lT_b)\mathrm{d}t = \frac{P_{r,1} T_b}{h\nu}. \tag{4.11}$$

The received waveform for the self-interference channel can be expressed as

$$h_i(t) = \int_{t - t_\mu}^{t} h_2(t)\mathrm{d}t = \frac{a}{p+1}(t - b)^{p+1} - \frac{a}{p+1}(t - t_\mu - b)^{p+1}. \tag{4.12}$$

Then $\lambda_{i,1}$ and $\lambda_{i,2}$ can be obtained as

$$\lambda_{i,1} = \frac{E_b a \left[(T_b - T_\tau - b)^{p+2} - (-b)^{p+2} - (T_b - T_\tau - t_\mu - b)^{p+2} + (-t_\mu - b)^{p+2} \right]}{h\nu(p+1)(p+2)},$$

$$\lambda_{i,2} = \frac{E_b a \left[(T_b - b)^{p+2} - (T_b - T_\tau - b)^{p+2} - (T_b - t_\mu - b)^{p+2} + (T_b - T_\tau - t_\mu - b)^{p+2} \right]}{h\nu(p+1)(p+2)}.$$

$$(4.13)$$

Besides, from (4.10) and from (4.13), we can define $\lambda_i \triangleq \lambda_{i,1} + \lambda_{i,2}$, which can be obtained as

$$\lambda_i = \frac{E_b}{h\nu} \int_0^{T_b} h_i(t)\mathrm{d}t = \frac{P_{r,2} T_b}{h\nu}. \tag{4.14}$$

Then the detected number of photons under the lth bit time can be expressed as

$$n = n_s x_{s,l} + n_{i,1} x_{i,l} + n_{i,2} x_{i,l-1} + n_b, \tag{4.15}$$

where n_s, $n_{i,1}$, $n_{i,2}$ and n_b are the received number of photons due to the information, the synchronous self-interference, the asynchronous self-interference and the background noise, respectively.

The distribution of n_s, $n_{i,1}$, $n_{i,2}$ and n_b can be assumed as Poisson distributions [9–11]. Because $x_{s,l}$, $x_{i,l}$, $x_{i,l-1} \in \{0, 1\}$, one can readily verify that n also follows a Poisson distribution with parameter $\lambda(x_{s,l}, x_{i,l}, x_{i,l-1})$ given in (4.8). Therefore, the probability density of the detected photon numbers n can be expressed as

$$p(n|x_{s,l}, x_{i,l}, x_{i,l-1})$$
$$= \frac{(x_{s,l}\lambda_s + x_{i,l}\lambda_{i,1} + x_{i,l-1}\lambda_{i,2} + \lambda_b)^n}{n!} e^{-(x_{s,l}\lambda_s + x_{i,l}\lambda_{i,1} + x_{i,l-1}\lambda_{i,2} + \lambda_b)}. \tag{4.16}$$

Then the probability of the detected photon numbers when bit 0 is transmitted can be obtained as

$$p(n|0) = \frac{1}{4} \sum_{x_{i,l}} \sum_{x_{i,l-1}} \frac{(x_{i,l}\lambda_{i,1} + x_{i,l-1}\lambda_{i,2} + \lambda_b)^n}{n!} e^{-(x_{i,l}\lambda_{i,1} + x_{i,l-1}\lambda_{i,2} + \lambda_b)}. \tag{4.17}$$

Similarly, the probability of the detected photon numbers when bit 1 is transmitted can be obtained as

$$p(n|1) = \frac{1}{4} \sum_{x_{i,l}} \sum_{x_{i,l-1}} \frac{(\lambda_s + x_{i,l}\lambda_{i,1} + x_{i,l-1}\lambda_{i,2} + \lambda_b)^n}{n!} e^{-(\lambda_s + x_{i,l}\lambda_{i,1} + x_{i,l-1}\lambda_{i,2} + \lambda_b)}. \tag{4.18}$$

We can assume that the information signal and the self-interference may arrive at the receiver synchronously when the asynchronous self-interference satisfies $\lambda_{i,1}/\lambda_i < 99.99\%$ [12]. The detected number of photons under the lth bit time when the information signal and the self-interference signal arrive at the receiver synchronously can be expressed as

$$n = n_s x_{s,l} + n_i x_{i,l} + n_b, \tag{4.19}$$

where n_i is the received number of photons due to the self-interference. Then the probability density of the detected photon numbers n for the synchronous receiver can be expressed as

$$p(n|x_{s,l}, x_{i,l}) = \frac{(x_{s,l}\lambda_s + x_{i,l}\lambda_i + \lambda_b)^n}{n!} e^{-(x_{s,l}\lambda_s + x_{i,l}\lambda_i + \lambda_b)}. \tag{4.20}$$

4.2 Bit-Error Rates and Achievable Information Rates

4.2.1 Bit-Error Rate and Achievable Information Rate for OOK

4.2.1.1 Bit-Error Rate for OOK

If we use the MAP detection to estimate the information bit $\hat{x}_{s,l}$ of the full-duplex UV communication, then the decision criterion for the MAP detection is given by

$$p(n|0) \underset{1}{\overset{0}{\gtrless}} p(n|1).$$

It is challenging to prove that this MAP detection for the full-duplex UV communication can reduce to a threshold detection. However, to obtain a tractable BER, we can still apply a threshold detection on the received number of photons and try to find the optimal threshold for minimizing the BER. In this case, the BER of the full-duplex UV communication can be expressed as

$$P_e = \frac{1}{2} \sum_{n=0}^{N_{th}} p(n|1) + \frac{1}{2} \sum_{n=N_{th}+1}^{\infty} p(n|0), \tag{4.21}$$

where the optimal threshold N_{th} can be obtained by numerical searching methods. Because $N_{th} \in [0, 1, \ldots, \infty]$, the numerical searching algorithm may suffer from low computational efficiency.

However, if we use an MAP detection to estimate the information bit $\hat{x}_{s,l}$ and the self-interference bits $\hat{x}_{i,l}$ and $\hat{x}_{i,l-1}$ simultaneously, then the decision criterion can be expressed as

$$\{\hat{x}_{s,l}, \hat{x}_{i,l}, \hat{x}_{i,l-1}\} = \arg\max_{x_{s,l},x_{i,l},x_{i,l-1}} p(n|x_{s,l}, x_{i,l}, x_{i,l-1}), \quad x_{s,l}, x_{i,l}, x_{i,l-1} \in \{0, 1\}. \tag{4.22}$$

To ensure an acceptable BER, we can assume that the signal strength is always larger than the self-interference, i.e., $\lambda_s > \lambda_i$, which is reasonable in practical implementations. Without loss of generality, we can further assume that $\lambda_i \geq \lambda_0$. Then we have the following theorem on the MAP detection.

The MAP detection defined in (4.22) for the OOK modulation of the full-duplex UV communication under self-interference is equivalent to a multiple-threshold rank detection, i.e.,

$$\{\hat{x}_{s,l}, \hat{x}_{i,l}, \hat{x}_{i,l-1}\} = \mathrm{Bin}(i) \quad \text{if} \quad N_{th}^i < n \leq N_{th}^{i+1}, \tag{4.23}$$

where $i = 0, 1, \ldots, 7$; $\mathrm{Bin}(i)$ is the binary representation of i; and the detection threshold N_{th}^i is given by

$$N_{th}^i = \begin{cases} 0, & i = 0; \\ \frac{\lambda(\mathrm{Bin}(i)) - \lambda(\mathrm{Bin}(i-1))}{\ln \lambda(\mathrm{Bin}(i)) - \ln \lambda(\mathrm{Bin}(i-1))}, & 1 \leq i \leq 7; \\ \infty, & i = 8. \end{cases} \tag{4.24}$$

If we focus on the information bit $\hat{x}_{s,l}$ only, we can use a single threshold detection

$$n \underset{1}{\overset{0}{\gtrless}} N_{th}^4 = \left\lfloor \frac{\lambda_s - \lambda_i}{\ln(\lambda_s + \lambda_b) - \ln(\lambda_i + \lambda_b)} \right\rfloor \tag{4.25}$$

to discriminate the information bit 0 and 1. In practical implementations, the background noise λ_b is usually small. Then the threshold can be approximated as

$\hat{N}_{\text{th}}^4 \approx \left\lfloor \dfrac{P_t T_b (10^{-\frac{PL_1}{10}} - 10^{-\frac{PL_2}{10}})}{(\frac{PL_2}{10} - \frac{PL_1}{10}) h\nu \ln 10} \right\rfloor$. We can observe that the threshold $\hat{N}_{\text{th}}^4$ of the full-duplex UV communication is approximately proportional to the transmitting power P_t. Equation (4.25) provides an analytical threshold for the threshold detection of the full-duplex UV communication, which is almost identical with the optimal threshold obtained by numerical searching methods in practical implementations.

Then we can obtain an approximated BER using the threshold given in (4.25) as

$$P_e = \frac{1}{2} + \frac{1}{8} \sum_{n=0}^{N_{\text{th}}^4} \sum_{x_{i,l}} \sum_{x_{i,l-1}} \left[\frac{(\lambda_s + x_{i,l}\lambda_{i,1} + x_{i,l-1}\lambda_{i,2} + \lambda_b)^n}{n!} e^{-(\lambda_s + x_{i,l}\lambda_{i,1} + x_{i,l-1}\lambda_{i,2} + \lambda_b)} \right.$$
$$\left. - \frac{(x_{i,l}\lambda_{i,1} + x_{i,l-1}\lambda_{i,2} + \lambda_b)^n}{n!} e^{-(x_{i,l}\lambda_{i,1} + x_{i,l-1}\lambda_{i,2} + \lambda_b)} \right].$$
$$(4.26)$$

We also present the BER when the information signal and the self-interference signal arrive at the receiver synchronously, which is a more simpler but useful case as we will demonstrate later in the numerical results. In this case, the BER can be simplified as

$$P_e = \frac{1}{2} + \frac{1}{4} \sum_{n=0}^{N_{\text{th}}^4} \frac{(\lambda_s + \lambda_b)^n e^{-(\lambda_s + \lambda_b)} + (\lambda_s + \lambda_i + \lambda_b)^n e^{-(\lambda_s + \lambda_i + \lambda_b)}}{n!}$$
$$(4.27)$$
$$- \frac{1}{4} \sum_{n=0}^{N_{\text{th}}^4} \frac{\lambda_b^n e^{-\lambda_b} + (\lambda_i + \lambda_b)^n e^{-(\lambda_i + \lambda_b)}}{n!}.$$

4.2.1.2   Achievable Information Rate for OOK

We assume that $\mathcal{X}_s$ is the input random process of the channel for the information signal; $\hat{\mathcal{X}}_s$ is the output random process of the channel. Then the AIR for the OOK modulation can be expressed as

$$R_I = K R_t I(\mathcal{X}_s; \hat{\mathcal{X}}_s) = K R_t \left[H(\hat{\mathcal{X}}_s) - H(\hat{\mathcal{X}}_s | \mathcal{X}_s) \right], \qquad (4.28)$$

where $R_t = \frac{1}{T_b}$ is the transmission bit rate; K is the number of communication links, where we have $K = 1$ and $K = 2$ for the half-duplex UV communication and the full-duplex UV communication, respectively; $I(\mathcal{X}_s; \hat{\mathcal{X}}_s)$ is the channel capacity; $H(\hat{\mathcal{X}}_s)$ and $H(\hat{\mathcal{X}}_s | \mathcal{X}_s)$ are respectively the entropy rate and the conditional entropy rate, which are defined as

$$H(\hat{\mathcal{X}}_s) \triangleq - \sum_{\hat{x}_s} p(\hat{x}_s) \log_2 p(\hat{x}_s) \qquad (4.29)$$

and

$$H(\hat{\mathcal{X}}_s | \mathcal{X}_s) \triangleq - \sum_{x_s} \sum_{\hat{x}_s} p(x_s) p(\hat{x}_s | x_s) \log_2 p(\hat{x}_s | x_s). \tag{4.30}$$

Here $p(\hat{x}_s) \triangleq \sum_{x_s} p(x_s) p(\hat{x}_s | x_s)$ is the posterior probability, where $p(x_s) = 0.5$ and $p(\hat{x}_s | x_s)$ is the conditional probability.

For the full-duplex UV communication, considering the impact of the self-interference and based on (4.17) and (4.18), the conditional probabilities $p(0|0)$ and $p(0|1)$ can be obtained as

$$p(0|0) = \sum_{n=0}^{N_{\text{th}}^4} p(n|0), \quad p(0|1) = \sum_{n=0}^{N_{\text{th}}^4} p(n|1). \tag{4.31}$$

4.2.2 *Bit-Error Rate and Achievable Information Rate for 4-PPM*

For the 4-PPM, each symbol f will be mapped into four transmission bits $\{x_m\}_{m=1}^4$. For a symbol $f \in \{1, 2, 3, 4\}$, we have $x_f = 1$ and the rest bits $x_m = 0$ with $m \neq f$. We denote the information symbol and the self-interference symbol by f_s and f_i, respectively. The lth transmitting bit x_l can be equivalently expressed as $x_{j,m}$, where j denotes the index of the symbol with $j \triangleq \lfloor \frac{l}{4} \rfloor$, and m is the bit index for the jth symbol with $m \triangleq \mod(l, 4)$.

4.2.2.1 Bit-Error Rate for 4-PPM

Now we focus on the detected number of photons $n_{j,m}$ for the mth bit in the jth symbol $x_{j,m}$ considering the impacts of the self-interference. When $m = 1$, the received photons will suffer the impact of asynchronous self-interference from the $(j-1)$th symbol; when $m \neq 1$, the received photons are only affected by the self-interference of the jth symbol. Therefore, the detected number of photons $n_{j,m}$ can be expressed as

$$n_{j,m} = \begin{cases} n_s x_{s,j,1} + n_{i,1} x_{i,j,1} + n_{i,2} x_{i,j-1,4} + n_b & \text{for } m = 1 \\ n_s x_{s,j,m} + n_{i,1} x_{i,j,m} + n_{i,2} x_{i,j,m-1} + n_b & \text{for } m \neq 1. \end{cases} \tag{4.32}$$

Then the probability density of the detected photon number $n_{j,m}$ can be obtained as

$$\begin{cases} p(n_{j,m}|x_{s,j,1}, x_{i,j,1}, x_{i,j-1,4}) \\ \quad = \dfrac{\lambda^n(x_{s,j,1}, x_{i,j,1}, x_{i,j-1,4})}{n!} e^{-\lambda(x_{s,j,1}, x_{i,j,1}, x_{i,j-1,4})} \quad \text{for} \quad m = 1 \\ p(n_{j,m}|x_{s,j,m}, x_{i,j,m}, x_{i,j,m-1}) \\ \quad = \dfrac{\lambda^n(x_{s,j,m}, x_{i,j,m}, x_{i,j,m-1})}{n!} e^{-\lambda(x_{s,j,m}, x_{i,j,m}, x_{i,j,m-1})} \quad \text{for} \quad m \neq 1, \end{cases} \tag{4.33}$$

where $\lambda(:,:,:)$ is the parameter of Poisson distribution given in (4.8). The BER of the 4-PPM for the full-duplex UV communication under self-interference can be obtained as

$$\begin{aligned}
P_e = 1 &- \frac{1}{64} \sum_{f_i=1}^{4} \sum_{n=0}^{\infty} p(n|1, x_{i,j,1}, x_{i,j-1,4}) \prod_{m=2}^{4} \left[\sum_{k=0}^{n} p(k|0, x_{i,j,m}, x_{i,j,m-1}) \right] \\
&- \frac{1}{64} \sum_{f_s=2}^{4} \sum_{f_i=1}^{4} \sum_{n=0}^{\infty} p(n|1, x_{i,j,f_s}, x_{i,j,f_s-1}) \left[\sum_{k=0}^{n} p(k|0, x_{i,j,1}, x_{i,j-1,4}) \right] \\
&\times \prod_{m \neq 1, m \neq f_s, m=2}^{4} \left[\sum_{k=0}^{n} p(k|0, x_{i,j,m,x_{i,j,m-1}}) \right].
\end{aligned} \tag{4.34}$$

When the information symbol and the self-interference symbol arrive at the receiver synchronously, the BER can be expressed as

$$\begin{aligned}
P_e = 1 &- \frac{1}{4} \sum_{n=0}^{\infty} \frac{(\lambda_s + \lambda_i + \lambda_b)^n e^{-(\lambda_s+\lambda_i+\lambda_b)}}{n!} \left(\sum_{k=0}^{n} \frac{\lambda_b^k}{k!} e^{-\lambda_b} \right)^3 \\
&- \frac{3}{4} \sum_{n=0}^{\infty} \frac{(\lambda_s + \lambda_b)^n e^{-(\lambda_s+\lambda_b)}}{n!} \left[\sum_{k=0}^{n} \frac{(\lambda_i + \lambda_b)^k e^{-(\lambda_i+\lambda_b)}}{k!} \right] \left(\sum_{k=0}^{n} \frac{\lambda_b^k}{k!} e^{-\lambda_b} \right)^2.
\end{aligned} \tag{4.35}$$

4.2.2.2 Achievable Information Rate for 4-PPM

The AIR for 4-PPM can be expressed as [14]

$$R_I = KJR_t I(\mathcal{X}_s; \hat{\mathcal{X}}_s) = KJR_t \left[H(\hat{\mathcal{X}}_s) - H(\hat{\mathcal{X}}_s|\mathcal{X}_s) \right], \tag{4.36}$$

where J is the modulation order with $J = 1/2$ for 4-PPM; $H(\hat{\mathcal{X}}_s)$ and $H(\hat{\mathcal{X}}_s|\mathcal{X}_s)$ are given by (4.29) and (4.30), respectively. The conditional probabilities can be expressed as

$$p(\hat{f}_{s,j} = 1 | f_s = 1) = \sum_{f_i=1}^{4} \sum_{n=0}^{\infty} p(n|1, x_{i,j,1}, x_{i,j-1,4}) \prod_{m=2}^{4} \left[\sum_{k=0}^{n} p(k|0, x_{i,j,m}, x_{i,j,m-1}) \right]$$

$$p(\hat{f}_{s,j} = 1 | f_s = 2, 3, 4) = \sum_{f_i=1}^{4} \sum_{n=0}^{\infty} p(n|0, x_{i,j,1}, x_{i,j-1,4}) \left[\sum_{k=0}^{n} p(k|1, x_{i,j,f_s}, x_{i,j,f_s-1}) \right]$$

$$\times \prod_{m\neq 1, m\neq f_s, m=2}^{4} \left[\sum_{k=0}^{n} p(k|0, x_{i,j,m}, x_{i,j,m-1}) \right].$$

(4.37)

Similar to (4.37), we can obtain the conditional probabilities $p(\hat{f}_{s,j} = 2 | f_s = 1, 2, 3, 4)$, $p(\hat{f}_{s,j} = 3 | f_s = 1, 2, 3, 4)$, and $p(\hat{f}_{s,j} = 4 | f_s = 1, 2, 3, 4)$.

4.3 Self-Interference Cancelling Method

We introduce the SIC method to mitigate the impacts of the self-interference, where we assume that the receiver of one user knows the transmitted bits of itself. In the SIC method, the receiver first estimates the average photon numbers of the self-interference when bit 1 is transmitted, then subtracts it from the received number of photons before making any decision.

4.3.1 SIC Method for OOK Modulation

4.3.1.1 Bit-Error Rate of SIC for OOK

The receiver knows the self-interference bits $\{x_{i,l}, x_{i,l-1}\}$ and their corresponding average number of photons $\{\lambda_{i,l}, \lambda_{i,l-1}\}$. Then after the SIC operation, the remaining number of photons can be expressed as

$$\hat{n} = n_s x_{s,l} + (n_{i,1} - \lambda_{i,1}) x_{i,l} + (n_{i,2} - \lambda_{i,2}) x_{i,l-1} + n_b. \tag{4.38}$$

Then the probability density of $\hat{n}$ given the information bit $x_{s,l}$ and the self-interference bits $x_{i,l}$ and $x_{i,l-1}$ can be obtained as

$$p(\hat{n} | x_{s,l}, x_{i,l}, x_{i,l-1}) = \frac{(x_{s,l}\lambda_s + x_{i,l}\lambda_{i,1} + x_{i,l-1}\lambda_{i,2} + \lambda_b)^{\hat{n}+\lambda_{i,1}x_{i,l}+\lambda_{i,2}x_{i,l-1}}}{(\hat{n} + \lambda_{i,1}x_{i,l} + \lambda_{i,2}x_{i,l-1})!}$$
$$\times e^{-(x_{s,l}\lambda_s + x_{i,l}\lambda_{i,1} + x_{i,l-1}\lambda_{i,2} + \lambda_b)}. \tag{4.39}$$

When the self-interference bits are "00", the number of photons $\hat{n}$ in (4.38) can be rewritten as $\hat{n} = n_s x_{s,l} + n_b$. Then an MAP detection with decision criteria $p(\hat{n}|0,0,0) \overset{0}{\underset{1}{\gtrless}} p(\hat{n}|1,0,0)$ is employed to discriminate the transmitted bits. Similar to the derivation of the Lemma 1, the MAP detection can be equivalently achieved by a threshold detection $\hat{n} \overset{0}{\underset{1}{\lessgtr}} N_{\text{th},1}$, where the threshold is given by $N_{\text{th},1} = \left\lfloor \frac{\lambda_s}{\ln(\lambda_s + \lambda_b) - \ln \lambda_b} \right\rfloor$. Similarly, when the self-interference bits are "01", "10" and "11", the corresponding thresholds can be obtained as $N_{\text{th},2} = \left\lfloor \frac{\lambda_s}{\ln(\lambda_s + \lambda_{i,2} + \lambda_b) - \ln(\lambda_{i,2} + \lambda_b)} \right\rfloor - \lambda_{i,2}$, $N_{\text{th},3} = \left\lfloor \frac{\lambda_s}{\ln(\lambda_s + \lambda_{i,1} + \lambda_b) - \ln(\lambda_{i,1} + \lambda_b)} \right\rfloor - \lambda_{i,1}$, and $N_{\text{th},4} = \left\lfloor \frac{\lambda_s}{\ln(\lambda_s + \lambda_i + \lambda_b) - \ln(\lambda_i + \lambda_b)} \right\rfloor - \lambda_i$, respectively. Given the thresholds $N_{\text{th},1}$, $N_{\text{th},2}$, $N_{\text{th},3}$, and $N_{\text{th},4}$ of the SIC detection, the BER can be obtained as

$$
\begin{aligned}
P_e = \frac{1}{8} \Big[& p(\hat{n} > N_{\text{th},1}|0,0,0) + p(\hat{n} > N_{\text{th},2}|0,0,1) + p(\hat{n} > N_{\text{th},3}|0,1,0) \\
& + p(\hat{n} > N_{\text{th},4}|0,1,1) + p(\hat{n} \le N_{\text{th},1}|1,0,0) + p(\hat{n} \le N_{\text{th},2}|1,0,1) \\
& + p(\hat{n} \le N_{\text{th},3}|1,1,0) + p(\hat{n} \le N_{\text{th},4}|1,1,1) \Big].
\end{aligned}
$$

(4.40)

When the information signal and the self-interference signal arrive at the receiver synchronously, the BER can be obtained as

$$
\begin{aligned}
P_e = \frac{1}{2} + \frac{1}{4} \sum_{\hat{n}=-\lambda_i}^{N_{\text{th},4}} & \frac{(\lambda_s + \lambda_i + \lambda_b)^{\hat{n}+\lambda_i} e^{-(\lambda_s + \lambda_i + \lambda_b)} - (\lambda_i + \lambda_b)^{\hat{n}+\lambda_i} e^{-(\lambda_i + \lambda_b)}}{(\hat{n} + \lambda_i)!} \\
& + \frac{1}{4} \sum_{\hat{n}=0}^{N_{\text{th},1}} \frac{(\lambda_s + \lambda_b)^{\hat{n}} e^{-(\lambda_s + \lambda_b)} - \lambda_b^{\hat{n}} e^{-\lambda_b}}{\hat{n}!}.
\end{aligned}
$$

(4.41)

4.3.1.2 Achievable Information Rate of SIC for OOK

The AIR of the SIC method for OOK modulation of the full-duplex UV communication can be obtained from (4.28). The conditional probabilities $p(0|0)$ and $p(0|1)$ can be respectively obtained as

$$
\begin{aligned}
p(0|0) = \frac{1}{4} [& p(\hat{n} \le N_{\text{th},1}|0,0,0) + p(\hat{n} \le N_{\text{th},2}|0,0,1) \\
& + p(\hat{n} \le N_{\text{th},3}|0,1,0) + p(\hat{n} \le N_{\text{th},4}|0,1,1)], \\
p(0|1) = \frac{1}{4} [& p(\hat{n} \le N_{\text{th},1}|1,0,0) + p(\hat{n} \le N_{\text{th},2}|1,0,1) \\
& + p(\hat{n} \le N_{\text{th},3}|1,1,0) + p(\hat{n} \le N_{\text{th},4}|1,1,1)].
\end{aligned}
$$

(4.42)

4.3.2   SIC Method for 4-PPM

4.3.2.1   Bit-Error Rate of SIC for 4-PPM

For the SIC method, both the self-interference bits $\{x_{i,l}, x_{i,l-1}\}$ and the corresponding average number of photons $\{\lambda_{i,l}, \lambda_{i,l-1}\}$ are perfectly known at the receiver. The number of photons for each transmitted bit after SIC operation can be expressed as

$$
\hat{n}_{j,m} =
\begin{cases}
n_s x_{s,j,1} + (n_{i,1} - \lambda_{i,1})x_{i,j,1} + (n_{i,2} - \lambda_{i,2})x_{i,j-1,4} + n_b, & \text{for } m = 1 \\
n_s x_{s,j,m} + (n_{i,1} - \lambda_{i,1})x_{i,j,m} + (n_{i,2} - \lambda_{i,2})x_{i,j,m-1} + n_b, & \text{for } m \neq 1.
\end{cases}
$$

(4.43)

The probability density of $\hat{n}_{j,m}$ can be obtained by (4.39). Then the BER of the SIC method for the full-duplex UV communication can be obtained as

$$
\begin{aligned}
P_e = 1 - \frac{1}{64} \sum_{f_i=1}^{4} \sum_{\hat{n}_{j,m}=-\lambda_{i,x,1}}^{\infty} p(\hat{n}_{j,m}|1, x_{i,j,1}, x_{i,j-1,4}) \prod_{m=2}^{4} \left[\sum_{k=-\lambda_{i,x,m}}^{\hat{n}_{j,m}} p(k|0, x_{i,j,m}, x_{i,j,m-1}) \right] \\
- \frac{1}{64} \sum_{f_s=2}^{4} \sum_{f_i=1}^{4} \sum_{\hat{n}_{j,m}=-\lambda_{i,x,f_s}}^{\infty} p(\hat{n}_{j,m}|1, x_{i,j,f_s}, x_{i,j,f_s-1}) \left[\sum_{k=-\lambda_{i,x,1}}^{\hat{n}_{j,m}} p(k|0, x_{i,j,1}, x_{i,j-1,4}) \right] \\
\times \prod_{m\neq 1, m\neq f_s, m=2}^{4} \left[\sum_{k=-\lambda_{i,x,m}}^{\hat{n}_{j,m}} p(k|0, x_{i,j,m}, x_{i,j,m-1}) \right].
\end{aligned}
$$

(4.44)

where $\lambda_{i,x,m}$ denotes the the average number of photons due to the self-interference with $\lambda_{i,x,m} \triangleq \lambda_{i,1} x_{i,j,m} + \lambda_{i,2} x_{i,j,m-1}$.

When the information signal and the self-interference signal arrive at the receiver synchronously, the BER can be obtained as

$$
\begin{aligned}
P_e = 1 - \frac{1}{4} \sum_{\hat{n}_{j,m}=-\lambda_i}^{\infty} \frac{(\lambda_s + \lambda_i + \lambda_b)^{\hat{n}_{j,m}+\lambda_i}}{(\hat{n}_{j,m} + \lambda_i)!} e^{-(\lambda_s+\lambda_i+\lambda_b)} \left[\sum_{k=0}^{\hat{n}_{j,m}} \frac{\lambda_b^k}{k!} e^{-\lambda_b} \right]^3 \\
- \frac{3}{4} \sum_{\hat{n}_{j,m}=0}^{\infty} \frac{(\lambda_s + \lambda_b)^{\hat{n}_{j,m}}}{\hat{n}_{j,m}!} e^{-(\lambda_s+\lambda_b)} \left[\sum_{k=-\lambda_i}^{\hat{n}_{j,m}} \frac{(\lambda_i + \lambda_b)^{k+\lambda_i} e^{-(\lambda_i+\lambda_b)}}{(k+\lambda_i)!} \right] \left[\sum_{k=0}^{\hat{n}_{j,m}} \frac{\lambda_b^k}{k!} e^{-\lambda_b} \right]^2.
\end{aligned}
$$

(4.45)

4.3.2.2 Achievable Information Rate of SIC for 4-PPM

The AIR for 4-PPM can be obtained from (4.36). The conditional probabilities of the SIC method for the full-duplex UV communication can be expressed as

$$p(\hat{f}_{s,j} = 1 | f_s = 1) = \sum_{f_i=1}^{4} \sum_{\hat{n}_{j,m}=-\lambda_{i,x,1}}^{\infty} p(\hat{n}_{j,m} | 1, x_{i,j,1}, x_{i,j-1,4})$$
$$\times \prod_{m=2}^{4} \left[\sum_{k=-\lambda_{i,x,m}}^{\hat{n}_{j,m}} p(k | 0, x_{i,j,m}, x_{i,j,m-1}) \right]$$

(4.46)

$$p(\hat{f}_{s,j} = 1 | f_s = 2, 3, 4) = \sum_{f_i=1}^{4} \sum_{\hat{n}_{j,m}=-\lambda_{i,x,1}}^{\infty} \left[\sum_{k=-\lambda_{i,x,f_s}}^{\hat{n}_{j,m}} p(k | 1, x_{i,j,f_s}, x_{i,j,f_s-1}) \right]$$
$$\times p(\hat{n}_{j,m} | 0, x_{i,j,1}, x_{i,j-1,4}) \prod_{m\neq 1, m\neq f_s, m=2}^{4} \left[\sum_{k=-\lambda_{i,x,m}}^{\hat{n}_{j,m}} p(k | 0, x_{i,j,m}, x_{i,j,m-1}) \right].$$

(4.47)

Similar to (4.37), we can obtain the conditional probabilities $p(\hat{f}_{s,j} = 2 | f_s = 1, 2, 3, 4)$, $p(\hat{f}_{s,j} = 3 | f_s = 1, 2, 3, 4)$, and $p(\hat{f}_{s,j} = 4 | f_s = 1, 2, 3, 4)$ accordingly.

4.4 Numerical Results

The path losses under different communication distances r and different elevation angles ϑ are shown in Fig. 4.3a and b, respectively. From Fig. 4.3, we can see that the path loss of the single scattering will gradually approach that of the two-order scattering and the path loss of the three-order scattering is almost one order of magnitude higher than that of the two-order scattering. Therefore, the influence of self-interferences mainly comes from the two-order scattering effect and the impact of scattering order higher than two can be ignored for NLOS full-duplex UV communications.

In Fig. 4.4, we present the BER performance of full-duplex UV communications versus the asynchronous rate for employing both the delta function and the rectangular function as the pulse shape. From Fig. 4.4a and b, we can observe that when the pulse width is small, the BER using rectangular function approaches that of using the delta function, which is as expected. Besides, the pulse width will change the asynchronous rate value corresponding to the minimum BER; and the minimum BER under different pulse width appears the same with each other. In Fig. 4.4a, we observe that the BER for OOK modulation fluctuates slightly as the asynchronous rate increases. In Fig. 4.4b, we see that the BER of the full-duplex communication for 4-PPM is sensitive to the asynchronous rate when the asynchronous rate is large.

Fig. 4.3 The path losses under different system geometries [15]

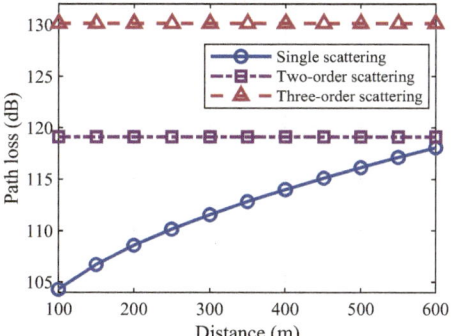

(a) The path losses with different communication distances

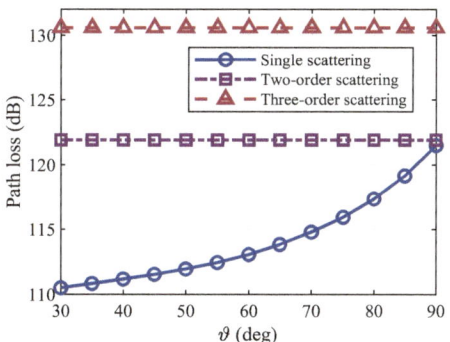

(b) the path losses with different elevation angles

From Fig. 4.4a and b, we also can see that the SIC method can greatly decreases the BER for the full-duplex UV communications. Besides, we can see that there is a slight fluctuation for the SIC method when the asynchronous rate is large. The reason is that the change rate of self-interferences increases as the asynchronous rate increases, especially when the asynchronous rate approaches 1. However, the curve of the BER for the SIC method is more stable as the asynchronous rate increases compared with the scheme without SIC method on the whole. Therefore, we can always approximate the BER using the one with asynchronous rate $\tau = 0$.

Figure 4.5 presents the BER performance versus the communication distance r and the elevation angle ϑ using OOK modulation. We employ a direct decision scheme as the comparison, where the self-interference is assumed unknown and the decision rule of the half-duplex scheme is used directly. From Fig. 4.5a and b, we can see that the BER of the OOK modulation for the full-duplex scheme is better

Fig. 4.4 The BER
performances of full-duplex
UV communications with
the asynchronous rate τ [15]

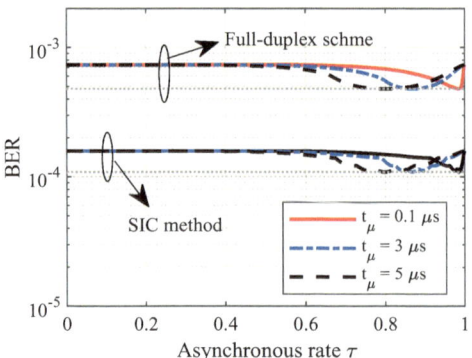

(a) OOK modulation

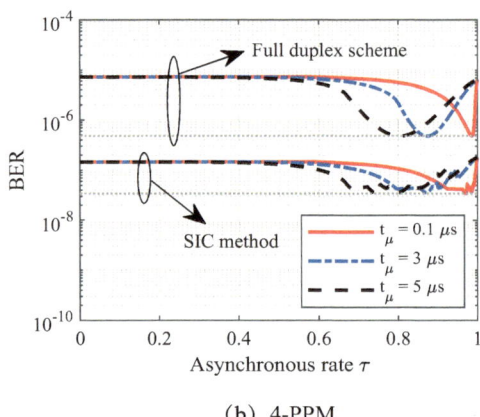

(b) 4-PPM

than that of the direct decision scheme, because we can obtain a better threshold
considering the impact of self-interferences. However, the SIC method can improve
the BER performance for the full-duplex scheme.

Figure 4.6 presents the AIR performance versus the communication distance r
and the elevation angle ϑ under different system geometries. From Fig. 4.6a, we can
observe that, when the communication distance is short, both the direct decision
scheme and the full-duplex scheme can achieve higher AIR than the half-duplex
scheme. However, these advantage over the half-duplex scheme will gradually diminish as the communication distance increases. Besides, by using the SIC method, the
full-duplex UV communication can hold the advantage over the half-duplex scheme
in wide range of communication distances. Similarly, from Fig. 4.6b, we can observe
that, when the elevation angles ϑ is small, both the direct decision scheme the full-
duplex scheme can achieve higher AIR than the half-duplex scheme. However, these

Fig. 4.5 The BER
performances of the
full-duplex UV
communication [15]

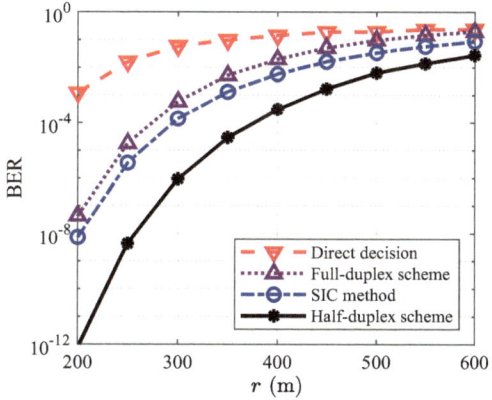

(a) The BER for OOK modulation
under different communication distances

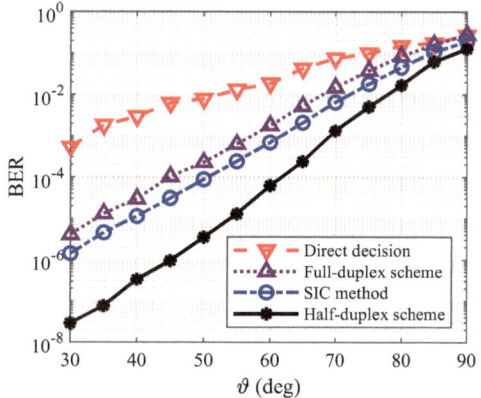

(b) The BER for OOK modulation
under different elevation angles

advantages over the half-duplex scheme will gradually diminish as the elevation
angles ϑ increases. Besides, by using the SIC method, the full-duplex UV communication can hold the advantage over the half-duplex scheme in wide range of elevation
angles.

Fig. 4.6 The AIR of the
full-duplex UV
communication under
different system geometries
[15]

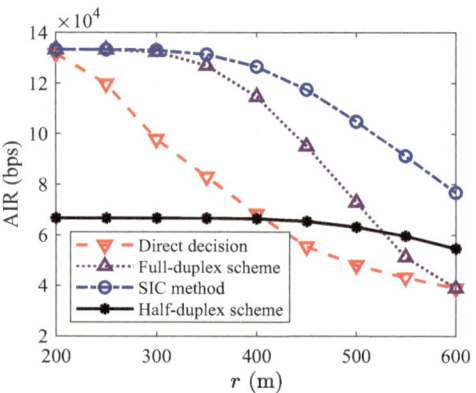

(a) The AIR under different communication distances

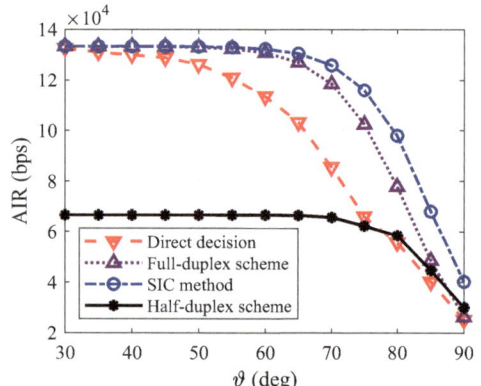

(b) The AIR under different elevation angles

4.5 Summary and Future Directions

In this chapter, we focus on the NLOS full-duplex UV communication in the presence of self-interferences. We introduced a power function model to estimate the CIR of the two-order scattering for NLOS UV communications. We also introduced the BER/PER and the AIR of OOK, 4-DPIM and 4-PPM for the NLOS full-duplex UV communication. Besides, we introduced the SIC method to mitigate the impact of self-interferences. The simulation results verified that the power function model can fit the CIRs of the two-order scattering accurately. Besides, we showed that the full-duplex UV communication can gradually lose its advantage over the half-duplex UV communication as either the distance or the elevation angle increases. Moreover, we presented that the full-duplex UV communication using the SIC method can hold its advantage over the half-duplex UV communication in a wide range of system geometries. This work clarified the application range of the full-duple UV communication and can provide some design guidelines for the full-duplex UV communication systems.

References

1. Yuan, R.Z., Ma, J.S., Su, P., et al.: Monte-Carlo integration models for multiple scattering based optical wireless communication. IEEE Tran. Commun. **68**(1), 334–348 (2020)
2. Siegel, A.M., Shaw, G.A., Model, J.: Short-range communication with ultraviolet LEDs. In: Fourth International Conference on Solid State Lighting, vol. 5530, pp. 182–193 (2004)
3. Xu, Z., Chen, G., Abou-Galala, F., et al.: Experimental performance evaluation of non-line-of-sight ultraviolet communication systems. In: Free-Space Laser Communications VII. SPIE, vol. 6709, pp. 287–298 (2007)
4. Chen, G., Abou-Galala, F., Xu, Z., et al.: Experimental evaluation of LED-based solar blind NLOS communication links. Opt. Express **16**(19), 15059–15068 (2008)
5. Chu, X., Yuan, R., Peng, M.: Turbulent single-scattering channel model for ultraviolet communications using equivalent scattering point approach. IEEE Commun. Lett. (2024)
6. Chen, G., Xu, Z., Sadler, B.M.: Experimental demonstration of ultraviolet pulse broadening in short-range non-line-of-sight communication channels. Opt. Express **18**(10), 10500–10509 (2010)
7. Han, D., Liu, Y., Zhang, K., et al.: Theoretical and experimental research on diversity reception technology in NLOS UV communication system. Opt. Express **20**(14), 15833–15842 (2012)
8. Deng, Y., Wang, Y., Zhang, Y.: The realization of a wide-angle voice transmission non-line-of-sight ultraviolet communication system. J. Semicond. **42**(9), 092301 (2021)
9. El-Shimy, M.A., Hranilovic, S.: On the use of photon arrival-times for non-line-of-sight solar-blind UV channels. IEEE Commun. Lett. **18**(6), 913–916 (2014)
10. Gong, C., Xu, Z.: Channel estimation and signal detection for optical wireless scattering communication with inter-symbol interference. IEEE Trans. Wirel. Commun. **14**(10), 5326–5337 (2015)
11. Wang, S., Peng, M., Yuan, R.: MIMO free-space optical communications using photon-counting receivers under weak links. IEEE Commun. Lett. **27**(4), 1185–1189 (2023)
12. Wang, Z., Yuan, R., Cheng, J., et al.: Joint optimization of full-duplex relay placement and transmit power for multi-hop ultraviolet communications. IEEE Internet Things J. (2023)
13. Guo, S., Park, K.H., Alouini, M.S.: Ordered sequence detection and barrier signal design for digital pulse interval modulation in optical wireless communications. IEEE Trans. Commun. **67**(4), 2880–2892 (2018)
14. Shannon, C.E.: ACM SIGMOBILE mobile computing and communications review. In: A Mathematical Theory of Communication, vol. 5, issue 1, pp. 3–55 (2001)
15. Wang, Z., Yuan, R., Peng, M.: Non-line-of-sight full-duplex ultraviolet communications under self-interference. IEEE Trans. Wirel. Commun. **22**(11), 7775–7788 (2023)

Chapter 5
Relay-Assisted Ultraviolet Communications

Abstract The full-duplex relay assisted ultraviolet (UV) communications can achieve longer communication distances and higher efficiency of time-frequency utilization compared with direct UV communications. However, due to the strong scattering effect, serious inter-relay-interference (IRI) is inevitably introduced in multi-hop full-duplex relay links. In this chapter, we introduce an alternate iterative-Newton method (AINM) for mitigating the impacts of IRI by optimizing jointly the relay placement and transmit powers of each relay. We first introduce the system model of relay-assisted multi-hop UV communication in Sect. 5.1. Then we derive the bit-error rate (BER) and the achievable information rate (AIR) of the consider communication system model in Sect. 5.2, where a space-division coupled full-duplex relay configuration is further proposed to mitigate the influence of IRI. In Sect. 5.3, we introduce the AINM to jointly optimizing the relay placement and the transmit power. In Sect. 5.4, we present some numerical results to explore the performance of relay-assisted UV communication systems. Numerical results show that the AINM can significantly decrease the BER and increase the AIR; and the space-division scheme can further decrease the BER but sacrifices about half AIR. At last, we conclude this chapter in Sect. 5.5.

Keywords Multi-hop communication · Relay-assisted communication · Relay placement

5.1 System Model of Relay Assisted Multi-hop Communications

Figure 5.1 presents the system configuration of the full-duplex relay assisted multi-hop UV communications. The system consists of one source node S, M relay nodes $\{R_1, R_2, \ldots, R_M\}$, and one destination node D. For simplicity, the source node and the destination node are also denoted by R_0 and R_{M+1}, respectively. We assume that all the nodes are sequenced in a line from the source to the destination and all the transmitting light beams and receiving field-of-views (FOV) are in coplanar

© The Author(s), under exclusive license to Springer Nature Singapore Pte Ltd. 2025   73
R. Yuan and Z. Wang, *Non-Line-of-Sight Ultraviolet Communications*,
SpringerBriefs in Computer Science, https://doi.org/10.1007/978-981-97-8543-8_5

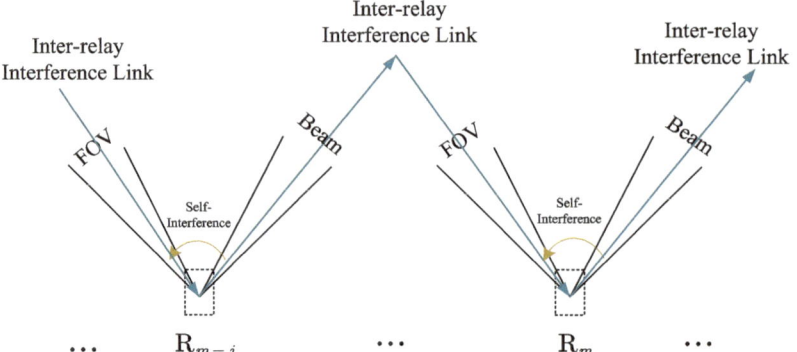

Fig. 5.1 System configuration of the full-duplex relay assisted multi-hop UV communications

geometry. The source S transmits an on-off keying (OOK) modulated signal to the first relay. Then the relay R_m decodes the signal from the relay R_{m-1} and forward the signal to the relay R_{m+1} with $m = 1, 2, \ldots, M$. Finally, the destination D decodes the signal from the last relay R_M. The maximum transmit power for each node is denoted by $P_{t,max}$ and the transmit power for the mth node is denoted by P_m. The distance between nodes R_i and R_j is denoted by $r_{i,j}$. Since all the relays are operating at the full-duplex mode, we assume that each relay simultaneously decodes the signal of tth time period and forward the signal of $(t-1)$th time period.

For a full-duplex relay assisted multi-hop UV communications, there are two types of interference need to be considered, i.e., the IRI between different relay links and the SI between the detection link and the forward link of each relay. The IRI is mainly due to the single-scattering effect and the two-order scattering effect when the link distance is long, which is modeled in the next subsection. The SI is dominated by the multiple scattering effects or potential reflecting effects and can vary in a large range [1]. Without loss of generality, we can regard the SI as a constant value since the relative geometries between the transmitter and the receiver at the same node assumed to be fixed. We further assume that the transmission rate is small enough such that the time spreading of the channel response is smaller than the bit interval; and therefore, the inter-symbol-interference between adjacent bits can be ignored. Then the channel can be simply characterized by the its path loss model.

In the following, we first present the channel path loss models of the single-scattering and the two-order scattering effects for NLOS UV communications, then derive the photon-counting properties and introduce the threshold detection of the signals, and finally derive the BER and the AIR of full-duplex relay assisted multi-hop UV communications.

5.1.1 Channel Path Loss Models

(1) Single-Scattering Model: The single-scattering model have been presented in Chap. 2. Here we briefly review the approximate path loss model used as the relay link model in this chapter. When the transmitting beam angle and the receiving FOV angle are small, the path loss due to the single-scattering effect can be approximated as [2]

$$L_1(r) = \frac{\alpha_s}{r} e^{-\beta_s r}, \tag{5.1}$$

where r is the distance between the transmitter and the receiver; the values of the parameters α_s and β_s depend on the transceiver geometries; and they are defined as [2]

$$\begin{cases} \alpha_s \triangleq \frac{A_r k_s p(\theta_s) \beta_T^2 \beta_R \sin\theta_s (12\sin^2\theta_R + \beta_R^2 \sin^2\theta_T)}{96\sin\theta_T \sin^2\theta_R (1-\cos(\beta_T/2))} \\ \beta_s \triangleq \frac{k_e(\sin\theta_T + \sin\theta_R)}{\sin\theta_s} \end{cases} \tag{5.2}$$

where A_r is the receiving area of the receiver; k_s and k_e are the scattering coefficient and the extinction coefficient of the UV light in the atmosphere, respectively; and we have $k_e = k_s + k_a$, where k_a is the absorption coefficient; θ_T and θ_R are the elevation angles of the transmitter and the receiver, respectively; $\theta_s = \theta_T + \theta_R$ is the scattering angle; β_T and β_R are the divergence angle of the transmitting beam and the FOV angle, respectively; $p(\theta_s)$ is the scattering phase function, which can be obtained as a linear combining of the Rayleigh scattering function $p^{Ray}(\theta_s)$ and the Mie scattering function $p^{Mie}(\theta_s)$, i.e., $p(\theta_s) = \frac{k_s^{Ray}}{k_s} p^{Ray}(\theta_s) + \frac{k_s^{Mie}}{k_s} p^{Mie}(\theta_s)$. The Rayleigh scattering function and the Mie scattering function are given by $p^{Ray}(\theta_s) = \frac{3[1+3\gamma+(1-\gamma)\cos^2\theta_s]}{16\pi(1+2\gamma)}$ and $p^{Mie}(\theta_s) = \frac{1-g^2}{4\pi} \left[\frac{1}{(1+g^2-2g\cos\theta_s)^{3/2}} + f\frac{0.5(3\cos^2\theta_s-1)}{(1+g^2)^{3/2}} \right]$, respectively, where γ, f, and g are model parameters [2].

(2) Two-Order Scattering Model: The path loss of two-order scattering is usually obtained by Monte-Carlo methods [3–6]. Similar to the single-scattering model, there is no general closed-form expression for the path loss of two-order scattering [7]. To facilitate our analysis, here we give a quadratic function to approximate the path loss of two-order scattering, i.e.,

$$L_2(r) = ar^2 + br + c, \tag{5.3}$$

where a, b, and c are fitting parameters obtained by fitting the quadratic function to the numerical results of the path losses obtained by the Monte-Carlo method.

5.1.2 Photon Counting Properties

Now we present the photon counting properties for the full-duplex relay assisted multi-hop UV communications. We consider the on-off keying modulation with equal prior probabilities and the DF relay scheme for each node. For the mth node, it can receive the signal from the previous $m - 1$ nodes and the SI from itself. Then the intensity of the received signal at the node R_m with $m = \{1, \ldots, M + 1\}$ given a set of transmitted bits $\boldsymbol{x}_m \triangleq \{x_0, \ldots, x_m\}$ can be expressed as

$$\lambda_m(\boldsymbol{x}_m) = \lambda_{m-1,m} x_{m-1} + \sum_{j=0}^{m-2} \lambda_{j,m} x_j + \lambda_{m,m} x_m + \lambda_b, \tag{5.4}$$

where $\lambda_{i,m} \triangleq \frac{P_i T_b}{h\nu} [L_1(r_{i,m}) + L_2(r_{i,m})]$ for $i \in \{0, 1, \ldots, m - 1\}$; $\lambda_{m,m} \triangleq \frac{P_m T_b g_r}{h\nu}$; g_r is the link gain of relay SI, and g_r belongs in $(g_{r,si}, \infty)$ due to scattering or reflection in the solar blind spectrum at the relay node [1]; $g_{r,si}$ is the SI due to two scattering link at the relay node. The first term $\lambda_{m-1,m} x_{m-1}$ is the signal; the second term $\sum_{j=0}^{m-2} \lambda_{j,m} x_j$ is the IRI; the third term $\lambda_{m,m} x_m$ is the SI; and the last term λ_b is the background noise; $x_m \in \{0, 1\}$ is the transmitted bit at the mth node; it satisfies $P_m \leq P_{t,max}$ and where $P_{t,max}$ is the maximum transmit power for each node.

Then the detected number of photons at the mth node with $m = \{1, \ldots, M + 1\}$ given a set of transmitted bits $\boldsymbol{x}_m$ can be modeled as a Poisson distribution with the density probability given by

$$p(n|\boldsymbol{x}_m) = \frac{[\lambda_m(\boldsymbol{x}_m)]^n}{n!} e^{-\lambda_m(\boldsymbol{x}_m)}, \tag{5.5}$$

We assume that the prior probabilities $p(x_0 = 0) = p(x_0 = 1) = \frac{1}{2}$; then when the error probability at each node is small, we can further approximate the prior probabilities of each relay node as $p(x_m = 0) \approx p(x_m = 1) \approx \frac{1}{2}$ for $m \in \{1, \ldots, M\}$. Then the probability of detecting n photons at the mth node when the transmitted bit at the $(m - 1)$th node is 0 can be obtained as

$$p(n|x_{m-1} = 0) = \sum_{\boldsymbol{x}_m | x_{m-1} = 0} \frac{\left[\lambda_m(\boldsymbol{x}_m | x_{m-1} = 0)\right]^n e^{-\lambda_m(\boldsymbol{x}_m | x_{m-1} = 0)}}{2^m n!}. \tag{5.6}$$

Similarly, the probability of detecting n photons at the mth node when the transmitted bit at the $(m - 1)$th node is 1 can be obtained as

$$p(n|x_{m-1} = 1) = \sum_{\boldsymbol{x}_m | x_{m-1} = 1} \frac{\left[\lambda_m(\boldsymbol{x}_m | x_{m-1} = 1)\right]^n e^{-\lambda_m(\boldsymbol{x}_m | x_{m-1} = 1)}}{2^m n!}. \tag{5.7}$$

5.2 Bit-Error Rates and Achievable Information Rates

5.2.1 Threshold Detection

(1) Full-Duplex Relay: The mth node needs to estimate the transmitted information bit at the $(m-1)$th node. If we use a maximum *a posteriori* probability (MAP) detection to estimate the information bit as $\hat{x}_{m-1}$, then the decision criterion for the MAP detection is given by $p(n|x_{m-1}=0) \overset{0}{\underset{1}{\gtrless}} p(n|x_{m-1}=1)$. However, it is challenging to prove that this MAP detection can reduce to a threshold detection. To obtain a tractable BER, we can still apply a threshold detection on the received number of photons and aim to find the optimal threshold at the mth node for minimizing the BER of the mth node. Besides, for the mth node, it knows the transmitted bit x_m of itself. Then we should use two different thresholds $N_{th,0}^m$ and $N_{th,1}^m$ for the cases with $x_m = 0$ and $x_m = 1$, respectively. Then the BER at the mth node for $x_m = 0$ and $x_m = 1$ can be respectively expressed as

$$
\begin{cases}
P_{e,m}(x_m = 0) = \frac{1}{2}\sum_{n=0}^{N_{th,0}^m} p(n|x_{m-1}=1, x_m = 0) \\
\quad + \frac{1}{2}\sum_{n=N_{th,0}^m+1}^{\infty} p(n|x_{m-1}=0, x_m = 0) \\
P_{e,m}(x_m = 1) = \frac{1}{2}\sum_{n=0}^{N_{th,1}^m} p(n|x_{m-1}=1, x_m = 1) \\
\quad + \frac{1}{2}\sum_{n=N_{th,1}^m+1}^{\infty} p(n|x_{m-1}=0, x_m = 1),
\end{cases}
\tag{5.8}
$$

where $p(n|x_{m-1} = i, x_m = j)$ is given by

$$
p(n|x_{m-1} = i, x_m = j) = \sum_{x_0,\dots,x_{m-2}} \frac{\left(\lambda_m^{i,j}\right)^n e^{-\lambda_m^{i,j}}}{2^{m-1}n!},
\tag{5.9}
$$

and where $\lambda_m^{i,j} \triangleq \lambda_m(x_m|x_{m-1} = i, x_m = j)$. The optimal thresholds $N_{th,0}^m$ and $N_{th,1}^m$ in (5.8) can only be obtained by numerical searching methods, which can be time-consuming. If we use a MAP detection to estimate the transmitted bits $\{x_0, \dots, x_{m-1}\}$ simultaneously, then the decision criterion for $\{\hat{x}_0, \dots, \hat{x}_{m-1}\}$ can be expressed as

$$
\{\hat{x}_0, \dots, \hat{x}_{m-1}\} = \arg\max_{x_0,\dots,x_{m-1}} p(n|x_0, \dots, x_{m-1}).
\tag{5.10}
$$

Because the path losses for both the single-scattering and the two-order scattering decrease as the distance increases, we have $\lambda_{0,m} < \lambda_{1,m} < \cdots < \lambda_{m-1,m}$. Then similar to the derivation in [8], the MAP detection in (5.10) is equivalent to a multiple-threshold rank detection; and if we only focus on the detection of the transmitted bit at the $(m-1)$th relay node, then the multiple-threshold rank detection will further degenerate to a single-threshold detection with the optimal threshold for $x_m = 0$ and $x_m = 1$ respectively given by [8]

$$N_{th,0}^m = \left\lfloor \left| \frac{\lambda_{m-1,m} - \sum_{j=1}^{m-1} \lambda_{j-1,m}}{\ln\left(\frac{\lambda_{m-1,m}+\lambda_b}{\sum_{j=1}^{m-1} \lambda_{j-1,m}+\lambda_b}\right)} \right| \right\rfloor \qquad (5.11)$$

and

$$N_{th,1}^m = \left\lfloor \left| \frac{\lambda_{m-1,m} - \sum_{j=1}^{m-1} \lambda_{j-1,m}}{\ln\left(\frac{\lambda_{m-1,m}+\lambda_{m,m}+\lambda_b}{\sum_{j=1}^{m-1} \lambda_{j-1,m}+\lambda_{m,m}+\lambda_b}\right)} \right| \right\rfloor . \qquad (5.12)$$

As we will demonstrate later in the numerical results, the derived optimal thresholds in (5.11) and (5.12) will coincide with the optimal thresholds obtained by minimizing the BER in (5.8).

(2) Space-Division Coupled Full-Duplex Relay: To reduce the impact of the IRI on full-duplex relay assisted multi-hop UV communications, we present the space-division-coupled full-duplex relay configuration, as shown in Fig. 5.2. The relay nodes with odd and even indexes respectively use different off-axis angles, ϕ_1 and ϕ_2, which ensures that the nodes with odd indexes and nodes with even indexes are in two different planes, i.e. Plane 1 an Plane 2 as shown in Fig. 5.2. Then the IRI between adjacent relay links is mitigated. Similar to the derivation for the full-duplex relay, the BERs $P_{e,m}(x_m = 0)$ and $P_{e,m}(x_m = 1)$ for the space-division coupled full-duplex relay configuration are in the same form with the full-duplex relay in (5.8), where the signal strength $\lambda_m(\boldsymbol{x}_m)$ is replaced with

$$\lambda_m(\boldsymbol{x}_m) = \lambda_{m-1,m}x_{m-1} + \sum_{j=1}^{\lfloor m/2 \rfloor} \lambda_{m-2j-1,m}x_{m-2j-1} + \lambda_{m,m}x_m + \lambda_b. \qquad (5.13)$$

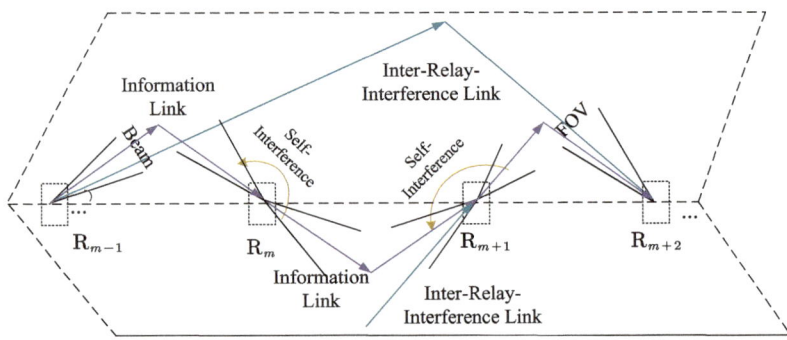

Fig. 5.2 Space-division coupled full-duplex relay configuration for multi-hop UV communications

Similarly, the optimal thresholds $N_{th,0}^m$ and $N_{th,1}^m$ can be respectively obtained as

$$N_{th,0}^m = \left\lfloor \frac{\lambda_{m-1,m} - \sum_{j=1}^{\lfloor m/2 \rfloor} \lambda_{m-2j-1,m}}{\ln \left(\frac{\lambda_{m-2j-1,m} + \lambda_b}{\sum_{j=1}^{\lfloor m/2 \rfloor} \lambda_{m-2j-1,m} + \lambda_b} \right)} \right\rfloor \qquad (5.14)$$

and

$$N_{th,1}^m = \left\lfloor \frac{\lambda_{m-1,m} - \sum_{j=1}^{\lfloor m/2 \rfloor} \lambda_{m-2j-1,m}}{\ln \left(\frac{\lambda_{m-1,m} + \lambda_{m,m} + \lambda_b}{\sum_{j=1}^{\lfloor m/2 \rfloor} \lambda_{m-2j-1,m} + \lambda_{m,m} + \lambda_b} \right)} \right\rfloor. \qquad (5.15)$$

(3) Half-Duplex Relay: Now we consider the half-duplex relay scheme for comparison. For a half-duplex relay scheme, each node transmits and detects the signal at different time slots. Therefore, there is no IRI and SI. Then the probabilities of detecting n photons when the transmitted bits of the previous node are 0 and 1 can be respectively expressed as

$$p_s(n|0) = \frac{\lambda_b^n}{n!} e^{-\lambda_b} \qquad (5.16)$$

and

$$p_s(n|1) = \frac{(\lambda_{m-1,m} + \lambda_b)^n}{n!} e^{-(\lambda_{m-1,m} + \lambda_b)}. \qquad (5.17)$$

When a MAP detection is employed, the optimal threshold at the mth node can be obtained as

$$N_{th}^m = \left\lfloor \frac{\lambda_{m-1,m}}{\ln(\lambda_{m-1,m} + \lambda_b) - \ln \lambda_b} \right\rfloor. \qquad (5.18)$$

5.2.2 Bit-Error Rate

(1) Full-Duplex Relay: After obtaining the decision thresholds $N_{th,0}^m$ in (5.11) and $N_{th,1}^m$ in (5.12), we can obtain the BER of the mth relay node for the full-duplex relay as

$$p_{e,m} = \frac{1}{2} P_{e,m}(x_m = 0) + \frac{1}{2} P_{e,m}(x_m = 1). \qquad (5.19)$$

By substituting (5.8) and (5.9) into (5.19), we can further rewrite the BER as

$$p_{e,m} = \frac{1}{2} + \sum_{x_0,\dots,x_{m-2}} \sum_{n=0}^{N_{th,0}^m} \frac{\left(\lambda_m^{1,0}\right)^n e^{-\lambda_m^{1,0}} - \left(\lambda_m^{0,0}\right)^n e^{-\lambda_m^{0,0}}}{2^{m+1} n!}$$

$$+ \sum_{x_0,\dots,x_{m-2}} \sum_{n=0}^{N_{th,1}^m} \frac{\left(\lambda_m^{1,1}\right)^n e^{-\lambda_m^{1,1}} - \left(\lambda_m^{0,1}\right)^n e^{-\lambda_m^{0,1}}}{2^{m+1} n!} \qquad (5.20)$$

Then the overall BER at the destination can be obtained as

$$p_e \approx 1 - \prod_{m=1}^{M+1} (1 - p_{e,m}) \approx \sum_{m=1}^{M+1} p_{e,m}, \qquad (5.21)$$

where we have assumed that $p_{e,m} \ll 1$.

(2) Space-Division Coupled Full-Duplex Relay: For the space-division coupled full-duplex relay, the BER at the mth relay node can similarly be obtained by substituting $\lambda_m(\boldsymbol{x}_m)$ of (5.13), $N_{th,0}^m$ of (5.14), and $N_{th,1}^m$ of (5.15) into (5.20); and the overall BER p_e at the destination can also be approximated as (5.21).

(3) Half-Duplex Relay: For the half-duplex relay, the BER at the mth relay node can be obtained as

$$p_{e,m} = \frac{1}{2} + \frac{1}{2} \sum_{n=0}^{N_{th}^m} \frac{(\lambda_{m-1,m} + \lambda_b)^n e^{-(\lambda_{m-1,m}+\lambda_b)} - \lambda_b^n e^{-\lambda_b}}{n!}. \qquad (5.22)$$

Similarly, the overall BER p_e at the destination can also be approximated as (5.21).

5.2.3 Achievable Information Rate

To derive the AIR of the system, we denote the random processes of transmitting information bits from the source node and receiving the bits at the destination by $\mathcal{X}_0$ and $\mathcal{X}_{M+1}$, respectively. Then the AIR of the relay assisted NLOS UV communication system can be expressed as

$$I = K_d I(\mathcal{X}_0; \mathcal{X}_{M+1}) = K_d \left[H(\mathcal{X}_0) - H(\mathcal{X}_{M+1}|\mathcal{X}_0) \right], \qquad (5.23)$$

where $K_d = \{1, \frac{1}{2}, \frac{1}{M+1}\}$ for the full-duplex relay, the space-division coupled full-duplex relay, and the half-duplex relay, respectively; $I(\mathcal{X}_0; \hat{\mathcal{X}}_M)$ is the channel capacity; $H(\mathcal{X}_0)$ and $H(\hat{\mathcal{X}}_M|\mathcal{X}_0)$ are the entropy and the conditional entropy given by

$$H(\mathcal{X}_0) \triangleq - \sum_{\hat{x}_M} p(\hat{x}_M) \log_2 p(\hat{x}_M) \qquad (5.24)$$

and

$$H(\hat{\mathcal{X}}_M|\mathcal{X}_0) \triangleq - \sum_{x_0} \sum_{\hat{x}_M} p(x_0) p(\hat{x}_M|x_0) \log_2 p(\hat{x}_M|x_0), \tag{5.25}$$

respectively, where $p(x_0)$ the prior probability of transmitting x_0 at the source node and we have $p(x_0 = 0) = p(x_0 = 1) = \frac{1}{2}$; $p(\hat{x}_M)$ is the posterior probability of the decoded bit x_M at the destination node with

$$p(\hat{x}_M) = \sum_{x_0} p(x_0) \prod_{m=1}^{M-1} p(x_{m+1}|x_m) p(\hat{x}_M|x_M), \tag{5.26}$$

where $p(\hat{x}_M|x_0) \triangleq \prod_{m=1}^{M-1} p(x_{m+1}|x_m) p(\hat{x}_M|x_M)$ is the conditional probability.

When the IRI is not serious, we can approximate channel between the source and the destination as a binary symmetrical channel with error probability given as p_e. Then we can approximate the AIR as

$$I \approx K_d[1 + p_e \log_2 p_e + (1 - p_e) \log_2(1 - p_e)]. \tag{5.27}$$

We will demonstrate later that the AIR given in (5.27) can well approximate the AIR given in (5.23).

5.3 Joint Optimization of Relay Placement and Transmit Power

5.3.1 Full-Duplex Relay

The change of either the relay placement or the relay transmit power will change the receiving number of photons $\lambda_{m-1,m}$ and further affect the BER performance of the full-duplex relay assisted UV communication systems. In this chapter, we achieve the minimum BER by jointly optimizing the placement and the transmit power of the relays. The optimization problem can be expressed as

$$\begin{aligned} \{\hat{r}, \hat{P}\} = \quad & \underset{r, P}{\arg\min} \quad p_e(r, P) \\ & \text{s.t.} \quad 0 < r_m < r_{m+1} < r, \\ & \qquad P_m \le P_{t,\max}, \end{aligned} \tag{5.28}$$

where $r \triangleq [r_1, r_2, \ldots, r_M]^T$ is an $M \times 1$ vector of the relay placement and r_m for $m \in \{1, 2, \ldots, M\}$ is the distance between the mth relay node and the source node; r is the distance between the source node and the destination node; $P \triangleq [P_0, P_1, \ldots, P_M]^T$

is an $(M + 1) \times 1$ vector of the transmit powers and P_m for $m \in \{0, 1, \dots, M\}$ is the transmit power of the mth node; $P_{t,\max}$ is the maximum transmit power of each relay.

Because the decision thresholds $N_{th,0}^m$ in (5.11) and $N_{th,1}^m$ in (5.12) are not derivable functions of either distance nor transmit power, it is challenging to solve the optimization problem in (5.28) using ordinary gradient descent methods. To solve the optimization problem, we present the AINM containing the following three steps:

- step 1 obtains the optimization results for $\boldsymbol{r}$ and $\boldsymbol{P}$ under fixed thresholds;
- step 2 updates the optimal thresholds according to (11) and (12) by using the obtained $\boldsymbol{r}$ and $\boldsymbol{P}$ in step 1;
- step 3 repeats step 1 and step 2 until the minimum BER is achieved.

The key of the AINM is to obtain the optimal $\boldsymbol{r}$ and $\boldsymbol{P}$ under fixed thresholds. However, it is difficult to prove that the optimizations (28) for both the transmit power $\boldsymbol{P}$ and the relay placement $\boldsymbol{r}$ are the convex problem under fixed thresholds. Then we use the alternating optimization method to solve this problem. Here we adopt the Newton's method to optimize $\boldsymbol{r}$ and $\boldsymbol{P}$ alternately. In the following, we present the optimization of the relay placement $\boldsymbol{r}$ given $\boldsymbol{P}$ and the optimization of the transmit power $\boldsymbol{P}$ given $\boldsymbol{r}$.

(1) Optimize $\boldsymbol{r}$ for Fixing $\boldsymbol{P}$: The Newton's method update the solution of the ith step using the gradient vector and the Hessian matrix of the objective function. Specifically, the relay placement is updated as

$$\boldsymbol{r}^i = \boldsymbol{r}^{i-1} - \frac{\nabla p_e(\boldsymbol{r})}{H(p_e(\boldsymbol{r}))}, \tag{5.29}$$

where the gradient vector $\nabla p_e(\boldsymbol{r})$ and the Hessian matrix $H(p_e(\boldsymbol{r}))$ are given by

$$\begin{cases} \nabla p_e(\boldsymbol{r}) \triangleq [\frac{\partial p_e}{\partial r_1}, \dots, \frac{\partial p_e}{\partial r_M}]^{\mathrm{T}} \\ H(p_e(\boldsymbol{r})) \triangleq \begin{bmatrix} \frac{\partial^2 p_e}{\partial r_1 \partial r_1} & \cdots & \frac{\partial^2 p_e}{\partial r_1 \partial r_M} \\ \cdots & \cdots & \cdots \\ \frac{\partial^2 p_e}{\partial r_M \partial r_1} & \cdots & \frac{\partial^2 p_e}{\partial r_M \partial r_M} \end{bmatrix}, \end{cases} \tag{5.30}$$

and where the first-order and the second-order partial derivatives $\frac{\partial p_e}{\partial r_k}$ and $\frac{\partial^2 p_e}{\partial r_k \partial r_l}$ for any $k, l \in \{1, 2, \dots, M\}$ can be expressed as

$$\begin{cases} \frac{\partial p_e}{\partial r_k} = \sum_{m=1}^{M+1} \frac{\partial p_{e,m}}{\partial r_k} \\ \frac{\partial^2 p_e}{\partial r_k \partial r_l} = \sum_{m=1}^{M+1} \frac{\partial^2 p_{e,m}}{\partial r_k \partial r_l}. \end{cases} \tag{5.31}$$

From (5.31) we can see that the key is to derive the partial derivatives $\frac{\partial p_{e,m}}{\partial r_k}$ and $\frac{\partial^2 p_e}{\partial r_k \partial r_l}$. Specifically, for $k > m$, we have $\frac{\partial p_{e,m}}{\partial r_k} = 0$; and for $k \leq m$, we have

$$\frac{\partial p_{e,m}}{\partial r_k} = \frac{1}{2^{m+1}} \sum_{x_0,\ldots,x_{m-2}} \sum_{n=0}^{N_{th,0}^m} \left(\frac{a_m^{1,0}}{n!} \frac{\partial \lambda_m^{1,0}}{\partial r_k} - \frac{a_m^{0,0}}{n!} \frac{\partial \lambda_m^{0,0}}{\partial r_k} \right)$$

$$+ \frac{1}{2^{m+1}} \sum_{x_0,\ldots,x_{m-2}} \sum_{n=0}^{N_{th,1}^m} \left(\frac{a_m^{1,1}}{n!} \frac{\partial \lambda_m^{1,1}}{\partial r_k} - \frac{a_m^{0,1}}{n!} \frac{\partial \lambda_m^{0,1}}{\partial r_k} \right) \qquad (5.32)$$

where $a_m^{i,j} \triangleq (\lambda_m^{i,j})^{n-1} e^{-\lambda_m^{i,j}} (n - \lambda_m^{i,j})$; and $\frac{\partial \lambda_m^{i,j}}{\partial r_k}$ is defined as

$$\frac{\partial \lambda_m^{i,j}}{\partial r_k} \triangleq \begin{cases} \sum_{p=0}^{k-1} \frac{x_p P_p T_b}{h\upsilon} \left[\frac{-\alpha_s e^{-\beta_s r_{p,k}} (\beta_s r_{p,k}+1)}{r_{p,k}^2} + 2ar_{p,k} + b \right], & k=m \\ \frac{x_k P_k T_b}{h\upsilon} \left[\frac{\alpha_s e^{-\beta_s r_{k,m}} (\beta_s r_{k,m}+1)}{r_{k,m}^2} - 2ar_{k,m} - b \right], & k < m. \end{cases} \qquad (5.33)$$

Based on the first-order derivative $\frac{\partial p_e}{\partial r_k}$, we can further derive the second-order derivative $\frac{\partial^2 p_e}{\partial r_k \partial r_l}$ from (5.31). Because the Hessian matrix is symmetric, we can focus on the cases with $l \leq k$ only. Then for $k > m$, we have $\frac{\partial^2 p_e}{\partial r_k \partial r_l} = 0$; and for $k \leq m$, we have

$$\frac{\partial^2 p_{e,m}}{\partial r_k \partial r_l} = \frac{1}{2^{m+1}} \sum_{x_0,\ldots,x_{m-2}} \left[\sum_{n=0}^{N_{th,0}^m} \left(\frac{b_m^{1,0} \partial^2 \lambda_m^{1,0} + a_m^{1,0} \partial^2 \lambda_m^{1,0}}{n! \partial r_k \partial r_l} \right) \right.$$

$$+ \sum_{n=0}^{N_{th,1}^m} \left(\frac{b_m^{1,1}}{n!} \frac{\partial^2 \lambda_m^{1,1}}{\partial r_k \partial r_l} + \frac{a_m^{1,1}}{n!} \frac{\partial^2 \lambda_m^{1,1}}{\partial r_k \partial r_l} \right)$$

$$- \sum_{n=0}^{N_{th,0}^m} \left(\frac{b_m^{0,0}}{n!} \frac{\partial^2 \lambda_m^{0,0}}{\partial r_k \partial r_l} + \frac{a_m^{0,0}}{n!} \frac{\partial^2 \lambda_m^{0,0}}{\partial r_k \partial r_l} \right) \qquad (5.34)$$

$$\left. - \sum_{n=0}^{N_{th,1}^m} \left(\frac{b_m^{0,1}}{n!} \frac{\partial^2 \lambda_m^{0,1}}{\partial r_k \partial r_l} + \frac{a_m^{0,1}}{n!} \frac{\partial^2 \lambda_m^{0,1}}{\partial r_k \partial r_l} \right) \right],$$

where $b_m^{i,j} \triangleq \left(\lambda_m^{i,j} \right)^{n-2} e^{-\lambda_m^{i,j}} \left[(n - \lambda_m^{i,j})^2 - n \right]$ and $\frac{\partial^2 \lambda_m^{i,j}}{\partial r_k \partial r_l}$ can be obtained as

$$\frac{\partial^2 \lambda_m^{i,j}}{\partial r_k \partial r_l} = \begin{cases} \sum_{p=0}^{l-1} \frac{x_p P_p T_b}{h\upsilon} \left[\frac{\alpha_s e^{-\beta_s r_{p,l}} [(\beta_s r_{p,l}+1)^2+1]}{r_{p,l}^3} + 2a \right], & l = k = m \\ -\frac{x_l P_l T_b}{h\upsilon} \left[\frac{\alpha_s e^{-\beta_s r_{l,k}} [(\beta_s r_{l,k}+1)^2+1]}{r_{l,k}^3} + 2a \right], & l < k = m \\ \frac{x_l P_l T_b}{h\upsilon} \left[\frac{\alpha_s e^{-\beta_s r_{l,m}} [(\beta_s r_{l,m}+1)^2+1]}{r_{l,m}^2} + 2a \right], & l = k < m \\ 0, & l < k < m. \end{cases} \qquad (5.35)$$

(2) *Optimize **P** for Fixing **r***: Then we optimize the transmit powers by fixing the relay placement. Similar to the optimization of the relay placement, the transmit powers are updated by

$$P^i = P^{i-1} - \frac{\nabla p_e(P)}{H(p_e(P))}, \tag{5.36}$$

where the gradient vector $\nabla p_e(P)$ and the Hessian matrix $H(p_e(P))$ are given by

$$
\begin{cases}
\nabla p_e(P) \triangleq [\frac{\partial p_e}{\partial P_0}, \ldots, \frac{\partial p_e}{\partial P_M}]^{\mathrm{T}} \\
H(p_e(P)) \triangleq \begin{bmatrix} \frac{\partial^2 p_e}{\partial P_0 \partial P_0} & \cdots & \frac{\partial^2 p_e}{\partial P_0 \partial P_M} \\ \cdots & \cdots & \cdots \\ \frac{\partial^2 p_e}{\partial P_M \partial P_0} & \cdots & \frac{\partial^2 p_e}{\partial P_M \partial P_M} \end{bmatrix},
\end{cases} \tag{5.37}
$$

and where the first-order and the second-order partial derivatives $\frac{\partial p_e}{\partial P_k}$ and $\frac{\partial^2 p_e}{\partial P_k \partial P_l}$ for any $k, l \in \{1, 2, \ldots, M\}$ can be expressed as

$$
\begin{cases}
\frac{\partial p_e}{\partial P_k} = \sum_{m=1}^{M+1} \frac{\partial p_{e,m}}{\partial P_k} \\
\frac{\partial^2 p_e}{\partial P_k \partial P_l} = \sum_{m=1}^{M+1} \frac{\partial^2 p_{e,m}}{\partial P_k \partial P_l}.
\end{cases} \tag{5.38}
$$

Here the key is to derive the partial derivatives $\frac{\partial p_{e,m}}{\partial P_k}$ and $\frac{\partial^2 p_e}{\partial P_k \partial P_l}$. Specifically, for $k \geq m$, we have $\frac{\partial p_{e,m}}{\partial P_k} = 0$; and for $k < m$, we have

$$
\begin{aligned}
\frac{\partial p_{e,m}}{\partial P_k} = & \frac{1}{2^{m+1}} \sum_{x_0, \ldots, x_{m-2}} \left(\sum_{n=0}^{N_{th,0}^m} \frac{a_m^{1,0} \partial \lambda_m^{1,0} - a_m^{0,0} \partial \lambda_m^{0,0}}{n! \partial P_k} \right. \\
& \left. + \sum_{n=0}^{N_{th,1}^m} \frac{a_m^{1,1} \partial \lambda_m^{1,1} - a_m^{0,1} \partial \lambda_m^{0,1}}{n! \partial P_k} \right),
\end{aligned} \tag{5.39}
$$

where

$$
\frac{\partial \lambda_m^{i,j}}{\partial P_k} \triangleq \begin{cases} \frac{x_m T_b g_r}{h\nu}, & k = m \\ \frac{x_k T_b}{h\nu}[L_1(r_{k,m}) + L_2(r_{k,m})], & k \neq m \end{cases} \tag{5.40}
$$

Based on the first-order derivative $\frac{\partial p_e}{\partial P_k}$, we can further derive the second-order derivative $\frac{\partial^2 p_e}{\partial P_k \partial P_l}$ from (5.38). Because the Hessian matrix is symmetrical, we can focus on the cases with $l \leq k$ only. Then for $k \geq m$, we have $\frac{\partial^2 p_{e,m}}{\partial P_k \partial P_l} = 0$; and for $k < m$, we have

$$
\begin{aligned}
\frac{\partial^2 p_{e,m}}{\partial P_k \partial P_l} = & \sum_{x_0, \ldots, x_{m-2}} \sum_{n=0}^{N_{th,0}^m} \frac{b_m^{1,0} \left(\partial \lambda_m^{1,0}\right)^2 - b_m^{0,0} \left(\partial \lambda_m^{0,0}\right)^2}{2^{m+1} n! \partial P_k \partial P_l} \\
& + \sum_{x_0, \ldots, x_{m-2}} \sum_{n=0}^{N_{th,1}^m} \frac{b_m^{1,1} \left(\partial \lambda_m^{1,1}\right)^2 - b_m^{0,1} \left(\partial \lambda_m^{0,1}\right)^2}{2^{m+1} n! \partial P_k \partial P_l}
\end{aligned} \tag{5.41}
$$

5.3.2 Space-Division Coupled Full-Duplex Relay

Similar to the joint optimization of r and P for the full-duplex relay, we can optimize r and P for the space-division coupled full-duplex relay by simply replacing $\lambda_m(x_m)$, $N_{th,0}^m$, and $N_{th,1}^m$ with (5.13), (5.14), and (5.15), respectively.

5.3.3 Half-Duplex Relay

For the half-duplex relay system, the BER $p_{e,m}$ of the mth relay node is given by (5.22), which depends on the transmit power P_{m-1} of the $(m-1)$th relay and the distance $r_{m-1,m}$ between the $(m-1)$th relay and the mth relay. The threshold N_{th}^m in $p_{e,m}$ is given by (5.18).

To facilitate our analysis, we define $f(\lambda_{m-1,m}) \triangleq \frac{(\lambda_{m-1,m}+\lambda_b)^n}{n!} e^{-(\lambda_{m-1,m}+\lambda_b)}$. Then the BER $p_{e,m}$ in (5.22) can be rewritten as

$$p_{e,m} = \frac{1}{2} + \frac{1}{2}\sum_{n=0}^{N_{th}^m}\left[f(\lambda_{m-1,m}) - \frac{\lambda_b^n}{n!}e^{-\lambda_b}\right]. \tag{5.42}$$

The partial derivative $\frac{\partial f(\lambda_{m-1,m})}{\partial \lambda_{m-1,m}}$ can be obtained as

$$
\begin{aligned}
&\frac{\partial f(\lambda_{m-1,m})}{\partial \lambda_{m-1,m}} \\
&= \frac{e^{-(\lambda_{m-1,m}+\lambda_b)}(\lambda_{m-1,m}+\lambda_b)^{n-1}}{n!}\left[n-(\lambda_{m-1,m}+\lambda_b)\right] \\
&< \frac{e^{-(\lambda_{m-1,m}+\lambda_b)}(\lambda_{m-1,m}+\lambda_b)^{n-1}}{n!}\left[\frac{\lambda_{m-1,m}}{\ln(\lambda_{m-1,m}+\lambda_b)-\ln\lambda_b}-\lambda_{m-1,m}\right] \\
&< 0.
\end{aligned}
\tag{5.43}
$$

Here in the last step we have assumed that $\lambda_{m-1,m} > (e-1)\lambda_b$, which is usually the case in practical systems. Since $\lambda_{m-1,m}$ is proportional to the transmit power P_{m-1}, then we can conclude that $p_{e,m}$ will decrease as the transmit power P_{m-1} increases for a fixing threshold N_{th}^m. However, the threshold N_{th}^m is also a function of the $\lambda_{m-1,m}$. Specifically, we can obtain the derivative of $N_{th}^m \lambda_{m-1,m}$ as

$$\frac{\partial N_{th}^m(\lambda_{m-1,m})}{\partial \lambda_{m-1,m}} = \frac{\ln\frac{\lambda_{m-1,m}+\lambda_b}{\lambda_b} - \frac{\lambda_{m-1,m}}{\lambda_{m-1,m}+\lambda_b}}{\left(\ln\frac{\lambda_{m-1,m}+\lambda_b}{\lambda_b}\right)^2}$$

$$> \frac{\ln\frac{\lambda_{m-1,m}+\lambda_b}{\lambda_b} - 1}{\left(\ln\frac{\lambda_{m-1,m}+\lambda_b}{\lambda_b}\right)^2} \tag{5.44}$$

$$> 0.$$

Here in the last step we have assumed that $\lambda_{m-1,m} > (e-1)\lambda_b$. Therefore, we can conclude that the threshold N_{th}^m will increase as the transmit power P_{m-1} increases. Then the number of summation terms in (5.42) will increases and the BER $p_{e,m}$ will decreases because $\frac{(\lambda_{m-1,m}+\lambda_b)^n}{n!}e^{-(\lambda_{m-1,m}+\lambda_b)} - \frac{\lambda_b^n}{n!}e^{-\lambda_b} < 0$ always holds for $n \le N_{th}^m$. Now we can finally conclude that the BER $p_{e,m}$ is a decreasing function of P_{m-1}. Therefore, to achieve the minimum overall BER of the half-duplex relay system, we should set the transmit power of each relay node as the maximum transmit power, i.e., $p_m = P_{t,max}$ for $m \in \{0, 1, \ldots, M\}$.

Then we consider the optimum relay placement for the half-duplex relay system. Similar to the analysis of the transmit power, we can readily find that the BER $p_{e,m}$ is a increasing function of $r_{m-1,m}$. Besides, we have proved that the transmit powers of all the relay nodes are the same with each other. Then the optimization of r is equivalent to

$$\hat{r} = \underset{r}{\arg\min} \sum_{m=1}^{M+1} p_{e,m}(r_{m-1,m}) \tag{5.45}$$

$$\text{s.t.} \quad r_{0,1} + r_{1,2} + \cdots + r_{M,M+1} = r.$$

Using the Lagrange multiplier method, one can readily find that the optimum solution is obtained when $r_{0,1} = r_{1,2} = \cdots = r_{M,M+1} = \frac{r}{M+1}$. Therefore, the minimum BER for half-duplex relay system is obtained when all relay nodes are placed at equal intervals.

5.4 Numerical Results

In this section, we present some numerical results to verify the performance of the joint optimization method. Unless otherwise specified, the simulation parameters are set as follows [9, 10]: the elevation angles of the transmitter and the receiver are $\theta_T = \theta_R = 45°$; the transmitting beam divergence angle $\beta_T = 17°$ and the receiving FOV angle $\beta_R = 30°$; the off-axis angles of the relay nodes with odd and even indexes for the space-division coupled full-duplex relay scheme are $\phi_1 = 45°$ and $\phi_2 = -45°$, respectively; the bit interval $T_b = 2\ \mu s$; the background noise $\lambda_b = 0.5$;

Fig. 5.3 The BER
performance under different
elevation angles [11]

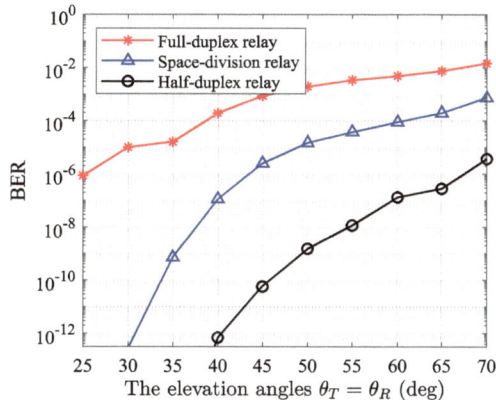

the distance between the source node and the destination node is $r = 500$ m; the
maximum transmit power is $P_{t,max} = 600$ mW; the searching precisions $\epsilon_r = 1$ m
and $\epsilon_P = 0.01$ mW.

We present the BER performance under different transceiver elevation angles for
the full-duplex relay, space-division coupled full-duplex relay, and the half-duplex
relay schemes in Fig. 5.3. The full-duplex relay without relay placement optimization is also plotted as a comparison scheme. The number of relays is $M = 3$ and
the maximum transmit power is $P_{t,max} = 800$ mW. From Fig. 5.3, we can see that
the BER increases as the elevation angle increases. This is because the path loss
between adjacent nodes increases as the elevation angle increases. The elevation
angle reflects the NLOS capability of UV communications. A higher elevation angle
indicates that UV communications can pass over higher obstacles and achieve better
NLOS capabilities. Because a smaller elevation angle corresponds a lower path loss,
the elevation angle for optimal performance is achieved at the smallest angle that
considers the limitation imposed by the obstacle. Besides, compared with the full-
duplex relay without relay placement optimization, the AINM can greatly improve
the BER performance of the full-duplex relay scheme. Moreover, from Fig. 5.3 we
can also see that the space-division coupled full-duplex relay scheme can further
improve the BER performance.

The corresponding relay placement results for the full-duplex relay scheme and the
space-division coupled full-duplex relay scheme are shown in Fig. 5.4. From Fig. 5.4,
we can see that the distance between adjacent nodes decreases as the index of relay
increases. Besides, we can also observe that the relay placement of space-division
coupled full-duplex relay scheme is more closer to the equal intervals compared with
the full-duplex relay scheme. This is because the full-duplex relay scheme suffers
more serious IRI than the space-division coupled full-duplex relay scheme, then
shorter communication distance should be allocated to the link with more serious
IRI to compensate the influence of IRI.

Fig. 5.4 The relay
placement under different
elevation angles [11]

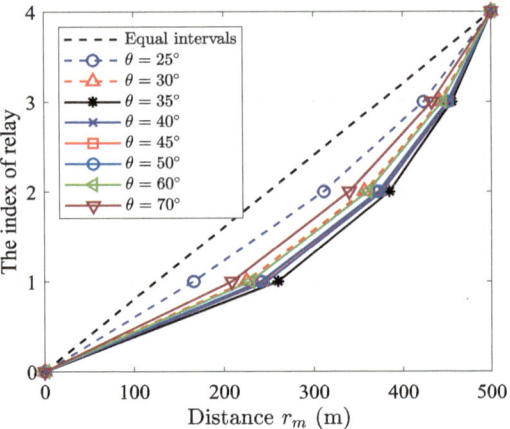

(a) The relay placement for full-duplex relay scheme

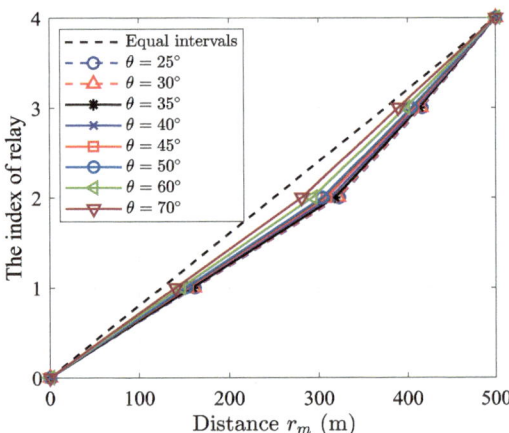

(b) The relay placement for space-division coupled full-
duplex relay scheme

The transmit power of each node under different elevation angles is shown in
Fig. 5.5. From Fig. 5.5, we can see that when the elevation angle is small, the transmit
power increases as the index of relay increases; while when the elevation angle is
large, the transmit power of each relay will approach the maximum transmit power.
This is because a small elevation angle corresponds to a strong communication link,
then a small transmit power is adequate to achieve the minimum BER and a larger
transmit power can result in larger IRI and thus worse BER performance; while a
larger elevation angle corresponds to a weak communication link, then the transmit
power becomes a scarce resource due to the high path loss; and therefore, each relay
node should adopt its maximum transmit power to achieve the minimum BER.

Fig. 5.5 The transmit power
of each relay under different
elevation angles [11]

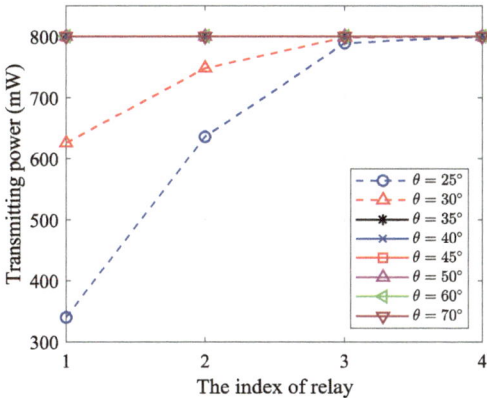

We then present the BER performance under different number of relay nodes for the full-duplex relay, space-division coupled full-duplex relay, and the half-duplex relay schemes in Fig. 5.6a. From Fig. 5.6a, we can see that the BER decreases as the number of relays increases for all the three relay schemes. However, the BER decreases slowly for the full-duplex relay scheme as the number of relays increases. This is because the IRI between adjacent relays increases as the number of relays increases. Then we present the AIR performance under different number of relays for the full-duplex relay, space-division coupled full-duplex relay, and the half-duplex relay schemes in Fig. 5.6b. From Fig. 5.6b we can see that for the full-duplex relay scheme, the AIR increases as the number of relays increases and gradually approaches a fixed value; while for the half-duplex scheme, the AIR decreases as the number of relays increases. This is because, for the full-duplex relay scheme, the BER becomes stable as the number of relays increases, which can be verified from the BER given in Fig. 5.6a; while for the half-duplex relay scheme, the transmitting rate will decreases as the number of relays increases.

5.5 Summary and Future Directions

The full-duplex relay assisted multi-hop UV communications will suffer from serious IRI due to the strong scattering effect of UV lights. In this chapter, we introduced the AINM to jointly optimize the relay placement and the transmit powers for full-duplex relay assisted multi-hop UV communications. Numerical results demonstrated that the AINM can greatly improve the BER performance of the full-duplex relay assisted multi-hop UV communications. Besides, numerical results showed that the space-division coupled full-duplex relay configuration can significantly improve the BER performance, but it sacrifices about half AIR.

Fig. 5.6 The BER and the
AIR under different number
of relays [11]

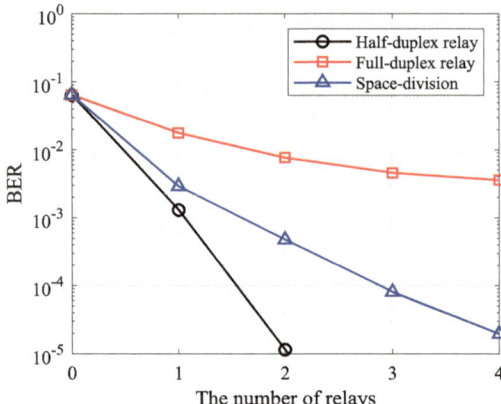

(a) The BER performance for multi-hop UV communication

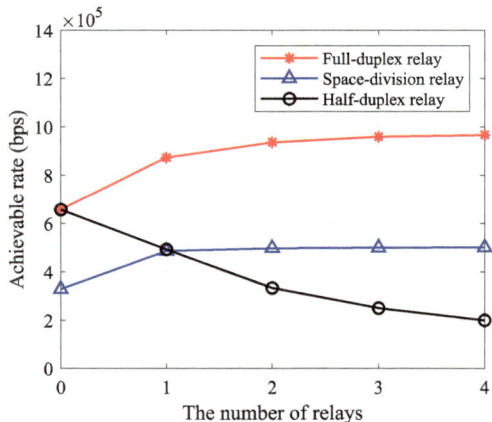

(b) The AIR performance for multi-hop UV communication

Previously, literature [12] studied the optimal relay placement for free-space optical communication in two-dimensional space, considering the constraints of both link obstacles and infeasible regions. However, the two-dimensional optimization problem is much complicated for UV communications due to the NLOS properties. In the future, we will extend our work for multi-hop UV communications to the two-dimensional space with link obstacles and infeasible regions.

References

1. Gong, C., Wang, K., Xu, Z., et al.: On full-duplex relaying for optical wireless scattering communication with on-off keying modulation. IEEE Trans. Wirel. Commun. **17**(4), 2525–2538 (2018)
2. Xu, Z., Ding, H., Sadler, B.M., et al.: Analytical performance study of solar blind non-line-of-sight ultraviolet short-range communication links. Opt. lett. **33**(16), 1860–1862 (2008)
3. Yuan, R., Ma, J., Su, P., et al.: Monte-Carlo integration models for multiple scattering based optical wireless communication. IEEE Trans. Commun. **68**(1), 334–348 (2019)
4. Ding, H., Chen, G., Majumdar, A.K., et al.: Modeling of non-line-of-sight ultraviolet scattering channels for communication. IEEE J. Sel. Areas Commun. **27**(9), 1535–1544 (2009)
5. Ding, H., Xu, Z., Sadler, B.M.: A path loss model for non-line-of-sight ultraviolet multiple scattering channels. EURASIP J. Wirel. Commun. Netw. 1–12 (2010)
6. Drost, R.J., Moore, T.J., Sadler, B.M.: UV communications channel modeling incorporating multiple scattering interactions. JOSA A **28**(4), 686–695 (2011)
7. Yuan, R., Ma, J., Su, P., et al.: An integral model of two-order and three-order scattering for non-line-of-sight ultraviolet communication in a narrow beam case. IEEE Commun. Lett. **20**(12), 2366–2369 (2016)
8. Wang, Z., Yuan, R., Peng, M.: Non-line-of-sight full-duplex ultraviolet communications under self-interference. IEEE Trans. Wirel. Commun. **22**(11), 7775–7788 (2023)
9. Chen, G., Abou-Galala, F., Xu, Z., et al.: Experimental evaluation of LED-based solar blind NLOS communication links. Opt. Express **16**(19), 15059–15068 (2008)
10. Chen, G., Xu, Z., Sadler, B.M.: Experimental demonstration of ultraviolet pulse broadening in short-range non-line-of-sight communication channels. Opt. Express **18**(10), 10500–10509 (2010)
11. Wang, Z., Yuan, R., Cheng, J., et al.: Joint optimization of full-duplex relay placement and transmit power for multi-hop ultraviolet communications. IEEE Internet Things J. (2023)
12. Zhu, B., Cheng, J., Alouini, M.S., et al.: Relay placement for FSO multihop DF systems with link obstacles and infeasible regions. IEEE Trans. Wirel. Commun. **14**(9), 5240–5250 (2015)

Chapter 6
Non-Line-of-Sight Ultraviolet Positioning

Abstract Optical positioning techniques have inherent advantages such as high precision, low power consuming and immunity to the electromagnetic interference. However, it is challenging to estimate the non-line-of-sight (NLOS) targets using current optical positioning methods. In this Chaper, we introduce the NLOS ultraviolet (UV) positioning method, which can obtain both the distance and the azimuth angle of the transmitter. We first review the early studies on UV positioning in Sect. 6.1. Then we introduce the NLOS UV positioning based on receiving powers in 6.2. Then we extend the UV positioning method to the full-duplex communication scenarios in Sect. 6.3. In Sect. 6.4 we present some numerical results of the NLOS UV Positioning. At last, we summarize this chapter and present some future directions on NLOS UV positioning in Sect. 6.5.

Keywords Ultraviolet positioning and Non-line-of-sight positioning · Ultraviolet signals

6.1 Early Study on Ultraviolet Positioning

Early exploration work mostly referred to the commonly used trilateral positioning method in wireless optical positioning technology or radio-frequency based positioning technology [1, 2]. For example, Ref. [3] studied the trilateral positioning method on a two-dimensional plane based on the single-scattering path loss model. The trilateral positioning method mainly uses the coordinates of three nodes on a two-dimensional plane to measure the received power and infer the position coordinates of the target node. As shown in Fig. 6.1, the positioning method demands a deployment of a certain proportion of anchor nodes in advance, equips them with GPS positioning systems, or obtains their own coordinates through some other special methods, then measures the distance between the anchor nodes and unknown target node, and finally calculates the coordinates of the target node through the trilateral positioning algorithm. Reference [4] extends the two-dimensional plane trilateral positioning method to the four-node positioning situation in a three-dimensional

© The Author(s), under exclusive license to Springer Nature Singapore Pte Ltd. 2025   93
R. Yuan and Z. Wang, *Non-Line-of-Sight Ultraviolet Communications*,
SpringerBriefs in Computer Science, https://doi.org/10.1007/978-981-97-8543-8_6

Fig. 6.1 Schematic diagram
of traditional triangulation
positioning

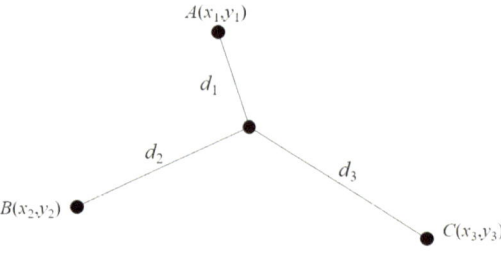

Fig. 6.2 Schematic diagram
of four-node positioning

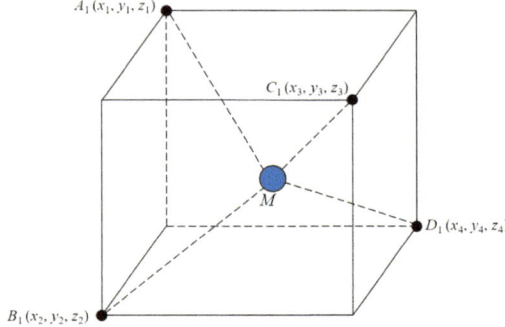

space, as shown in Fig. 6.2. In this three-dimensional positioning, the coordinates of
the four anchor nodes and the distance information from the unknown node to the
four anchor nodes need to be known, and the three-dimensional coordinates of the
unknown node can be calculated.

However, the above studies all assume that the receiver has accurate transmission
elevation information and requires multiple nodes to collaborate and determine the
target node position. In practical systems, it is challenging to apply the trilateral positioning method or the four-node positioning method before the communication links
between multiple nodes are established. Therefore, the positioning method needs to
be completed by a single node. The single-node positioning method first appeared
in the research of neighboring node search technology for UV communication networks. For example, Ref. [5] proposed a method to determine the orientation of the
transmitter by measuring the received light intensity. Reference [5] utilizes channel
quality that varies along different scattering directions to sort the node pointing directions between each pair of nodes, achieving efficient neighbor node search. Based on
this method, Refs. [6, 7] respectively proposed neighboring node search techniques
by using guidance node allocation protocol and credit collection protocol, which
greatly improves the efficiency of neighboring node search algorithms. Reference
[6] reduces the collision probability based on the random access method by selecting
a leader for the arbitration search process. Reference [7] proposes a more effective
neighboring node feedback mechanism to reduce the running time of credit collection protocols and improve the efficiency of neighboring node search. Reference [8]

proposes a more accurate transmitter direction sensing method, which requires the transmitter to point towards the receiver and estimate the incident angle of the transmitter by comparing the received signal strengths of two adjacent antennas with the same elevation angle in an omnidirectional antenna array. In addition, Ref. [8] used a photomultiplier tube as the receiver to experimentally verify the effectiveness of the neighboring node search protocol, and found that under typical geometric structures, its directional positioning accuracy can reach up to $3\,°C$ or less over short distances.

However, the neighbor node search technique studied in Refs. [5–8] requires strict limitations on the geometric parameters of the transceiver. For example, Refs. [5–7] require all transceivers to use the same transceiver geometry, while Ref. [8] requires the transmitter direction to always align with the receiver. These requirements are often not achievable in practical systems. In addition, Refs. [5–8] can only obtain a rough orientation of the transmitter, and cannot obtain the distance between the transmitter and itself at all. In practical UV communication scenarios, receivers usually cannot obtain prior information on the direction of the transmitter, making it difficult to apply above neighbor node search technology to accurately locate the UV communication nodes.

From the above research status, it can be seen that existing NLOS UV positioning methods are limited by the emission direction and cannot estimate the distance of the transmitter. In practical applications, the transmitter can point in any direction, but the receiver is usually lack of prior information about the pointing direction of the transmitter. In this chapter, we introduce the single-node based NLOS UV positioning methods suitable for arbitrary transmitter geometry.

6.2 NLOS Ultraviolet Positioning Based on Receiving Powers

The main purpose of UV positioning is to obtain the location of transmitter [11]. Here we assume the receiver knows the transmitting power P_T and the divergence angle β_T of the transmitting beam, which is usually the case when the transmitter and the receiver belong to two friendly users. A typical scenario of the NLOS UV positioning using two photon-counting receivers is shown in Fig. 6.3. The two receivers (Rx) are located at the origin of coordinate system C (x, y, z); $l_{R,i}$ is the ray of the ith receiving FOV axis, where $i \in \{1, 2\}$; l is the ray from the receiver to the unknown transmitter (Tx) with direction vector μ_l; ϕ_l is the azimuth angle between l and x-axis; r is the distance between receiver and transmitter; l_T denotes the transmitting beam axis with direction vector μ_T. The plane α is determined by μ_T and μ_l. Then the UV positioning is equivalent to obtaining the unknown geometry parameters ϕ_l, r, θ_T and ϕ_T.

The proposed UV positioning method can be divided into two steps. First, we estimate the azimuth angle ϕ_l by intersecting the ground plane and the plane α

Fig. 6.3 Scenario of NLOS
UV positioning by using two
receivers [11]

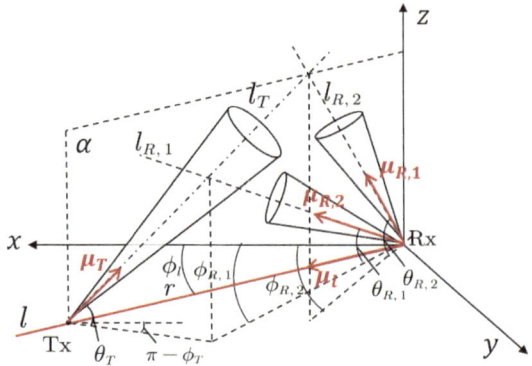

formed by the two receiving axes and passing through the transmitting beam axis.
Second, we estimate the distance and pointing direction of the transmitter by using
the single-scattering model.

6.2.1   The Azimuth Angle of the Transmitter

The key of obtaining the azimuth angle of the transmitter is to obtain the plane α,
which can be estimated by two rays $l_{R,1}$ and $l_{R,2}$ intersecting with l_T. Because the
maximum receiving power always happens when the transmitting beam axis and
the receiving FOV axis are in coplanar geometry for the single-scattering model
with narrow beam and receiving FOV, we can find rays $l_{R,1}$ and $l_{R,2}$ by rotating the
two receiver around the z-axis and obtaining the maximum receiving powers at two
different elevation angles.

For the ith receiver, we rotate it around the z-axis and measure the receiving powers
$\{P_{R,i}^1(\phi^1), P_{R,i}^2(\phi^2), \ldots, P_{R,i}^M(\phi^M)\}$ at M different azimuth angles $\{\phi^1, \phi^2, \ldots, \phi^M\}$.
Then the maximum receiving power and the corresponding azimuth angle can be
obtained as

$$
\begin{cases}
P_{R,i} = \max\{P_{R,i}^1(\phi^1), P_{R,i}^2(\phi^2), \ldots, P_{R,i}^M(\phi^M)\} \\
\phi_{R,i} = \underset{\{\phi^1, \ldots, \phi^M\}}{\arg\max}\{P_{R,i}^1(\phi^1), P_{R,i}^2(\phi^2), \ldots, P_{R,i}^M(\phi^M)\}.
\end{cases}
\tag{6.1}
$$

The photon-counting receiver can obtain the receiving power $P_{R,i}^m$ by estimating
the average number of photons as

$$
\hat{P}_{R,i}^m = \frac{\lambda_{s,i}h\nu}{T_d},
\tag{6.2}
$$

where h is the plank constant; v is the frequency of the transmitted UV light; $\lambda_{s,i}$ is the average number of photons, which can be obtained by averaging the number of received photons in K bit intervals according to the law of large numbers, i.e.,

$$\lambda_{s,i} = \frac{1}{K} \sum_{k=1}^{K} n_k. \tag{6.3}$$

Here n_k is the detected number of photon at the kth bit interval satisfying the Poisson distribution with parameter $\lambda_{s,i}$, i.e.,

$$p(n_i) = \frac{\lambda_{s,i}^{n_i}}{n_i!} e^{-\lambda_{s,i}}. \tag{6.4}$$

According to (6.23), the estimated $\hat{\phi}_{R,i}$ is obtained as the corresponding azimuth angle of $\hat{P}_{R,i}$. In practical implementations, the receiving azimuth angle is directly read by a protractor or a goniometer; and thus the distribution of $\hat{\phi}_{R,i}$ is closely related to the employed instruments. For simplicity, we assume that the receiving azimuth angle $\hat{\phi}_{R,i}$ satisfies a Gaussian distribution with mean $\phi_{R,i}$ and variance $\sigma_{\hat{\phi}_{R,i}}^2$.

After obtaining the azimuth angles $\phi_{R,1}$ and $\phi_{R,2}$, we can express the direction vectors of the ray $l_{R,1}$ and $l_{R,2}$ as

$$\begin{cases} \mu_{R,1} = \left[\cos\theta_{R,1}\cos\phi_{R,1}, \cos\theta_{R,1}\sin\phi_{R,1}, \sin\theta_{R,1}\right]^{\mathrm{T}} \\ \mu_{R,2} = \left[\cos\theta_{R,2}\cos\phi_{R,2}, \cos\theta_{R,2}\sin\phi_{R,2}, \sin\theta_{R,2}\right]^{\mathrm{T}}. \end{cases} \tag{6.5}$$

where $\theta_{R,1}, \theta_{R,2}$ are the elevation angles of $l_{R,1}$ and $l_{R,2}$, respectively. Then the plane α containing the transmitter and the receiver is formed by $\mu_{R,1}$ and $\mu_{R,2}$; and the norm vector of the plane α can be obtained by the cross product of $\mu_{R,1}$ and $\mu_{R,2}$

$$\mu_\alpha = \mu_{R,1} \times \mu_{R,2} \triangleq [a, b, c]^{\mathrm{T}}, \tag{6.6}$$

where a, b, and c are defined as

$$\begin{cases} a \triangleq \cos\theta_{R,1}\sin\phi_{R,1}\sin\theta_{R,2} - \cos\theta_{R,2}\sin\phi_{R,2}\sin\theta_{R,1} \\ b \triangleq -\cos\theta_{R,1}\cos\phi_{R,1}\sin\theta_{R,2} + \cos\theta_{R,2}\cos\phi_{R,2}\sin\theta_{R,1} \\ c \triangleq \cos\theta_{R,1}\cos\theta_{R,2}\sin\left(\phi_{R,2} - \phi_{R,1}\right). \end{cases} \tag{6.7}$$

Then the equation of plane α can be expressed as

$$\alpha : ax + by + cz = 0 \tag{6.8}$$

The intersection of plane α and the ground plane x-o-y is line l and the equation of l can be expressed as

$$l : \begin{cases} ax + by + cz = 0 \\ z = 0. \end{cases} \tag{6.9}$$

The slop of line l in plane x-o-y can be obtained by $\tan \phi_l$. Then azimuth angle ϕ_l can be expressed as

$$\phi_l = \arctan \left(-\frac{a}{b} \right) + k\pi, \tag{6.10}$$

where $k \in \mathbb{Z}$. When ϕ_l is restricted in $[0, 2\pi]$, there are two azimuth angle ϕ_l^1 and ϕ_l^2 satisfy (6.10), which are at inverse directions because $|\phi_l^1 - \phi_l^2| = \pi$. Nevertheless, we choose the one with larger receiving power, which corresponds to the one with $\boldsymbol{\mu}_{R,i} \cdot \boldsymbol{\mu}_l > 0$, and where $\boldsymbol{\mu}_l \triangleq [\cos \phi_l, \sin \phi_l, 0]^{\mathrm{T}}$ is the direction vector of ray l.

6.2.2 Distance and Pointing Direction of Beam Axis

The distance and pointing direction of the transmitter can be estimated by the coplanar single-scattering model at plane α. The typical geometry setting for the coplanar single-scattering model of UV communication system is shown in Fig. 6.4: r is the distance between Tx and Rx; θ_t and θ_r are the elevation angles of Tx and Rx, respectively; β_t is the divergence angle of the light beam at Tx; β_r is the FOV angle of Rx. The receiving power of the NLOS UV communication can be estimated by the single-scattering model in short range communications. Besides, the average receiving power of the single-scattering in a coplanar geometry with narrow beam and receiving FOV can be approximated as [9, Eq. (5)]

Fig. 6.4 Coplanar
single-scattering model of
UV communication system

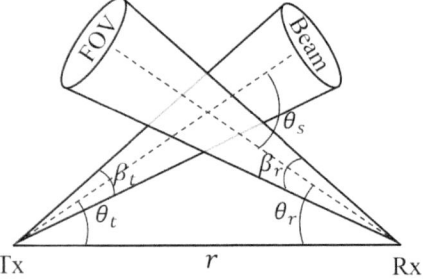

$$P_R(r, \theta_t, \theta_r) = \frac{P_T A_r k_s \beta_t^2 \beta_r P(\cos\theta_s) \sin\theta_s \left(12\sin^2\theta_r + \beta_r^2 \sin^2\theta_t\right)}{96r \sin\theta_t \sin^2\theta_r \left(1 - \cos\dfrac{\beta_t}{2}\right) \exp\left(\dfrac{k_e r(\sin\theta_t + \sin\theta_r)}{\sin\theta_s}\right)},$$

(6.11)

where P_T is transmitted light power; A_r is the receiving area; k_s is the scattering coefficient; k_e is the extinction coefficient; $\theta_s = \theta_t + \theta_r$ is the scattering angle; $P(\cos\theta_s)$ is the scattering phase function, which can be obtained as a linear combination of the Rayleigh scattering function $P^{Ray}(\cos\theta_s)$ and the Mie scattering function $P^{Mie}(\cos\theta_s)$ [9, Eq. (2)], i.e.,

$$P(\cos\theta_s) = \frac{k_s^{Ray}}{k_s} P^{Ray}(\cos\theta_s) + \frac{k_s^{Mie}}{k_s} P^{Mie}(\cos\theta_s)$$

(6.12)

where k_s^{Ray} is the Rayleigh scattering coefficient; k_s^{Mie} is the Mie scattering coefficient. The Rayleigh scattering function and Mie scattering function can be expressed as [9, Eqs. (3)–(4)]

$$P^{Ray}(\cos\theta_s) = \frac{3\left[1 + 3\gamma + (1 - \gamma)\cos^2\theta_s\right]}{16\pi(1 + 2\gamma)}$$

(6.13)

and

$$P^{Mi.e.}(\cos\theta_s) = \frac{1 - g^2}{4\pi}\left[\frac{1}{\left(1 + g^2 - 2g\cos\theta_s\right)^{3/2}} + f\frac{0.5\left(3\cos^2\theta_s - 1\right)}{\left(1 + g^2\right)^{3/2}}\right],$$

(6.14)

respectively, where γ, f and g are model parameters.

The coplanar single-scattering model has two unknown parameters r and θ_t. The elevation angles of receiver $\theta_{r,1}$ and $\theta_{r,2}$ can be obtained by the arccosine of dot product $s_{r,i} \cdot s_l$, i.e.,

$$\begin{cases} \theta_{r,1} = \arccos\left(\cos\theta_{R,1}\cos\left(\phi_{R,1} - \phi_l\right)\right) \\ \theta_{r,2} = \arccos\left(\cos\theta_{R,2}\cos\left(\phi_{R,2} - \phi_l\right)\right). \end{cases}$$

(6.15)

Then the distance r and the elevation angle θ_t of the beam axis in plane α can be obtained by solving the following equations

$$\begin{cases} P_R\left(r, \theta_t, \theta_{r,1}\right) = P_{R,1} \\ P_R\left(r, \theta_t, \theta_{r,2}\right) = P_{R,2}. \end{cases}$$

(6.16)

The equations in (6.16) can be solved by finding the solution to the following optimization problem:

$$\left\{\hat{r}, \hat{\theta}_t\right\} = \underset{\{r,\theta_t\}}{\arg\min} \sqrt{\sum_{i=1}^{2} \left[P_r\left(r, \theta_t, \theta_{r,i}\right) - P_i\right]^2}$$

$$\text{s.t.} \quad 0 < r < r_{max}$$
$$0 < \theta_t < \pi,$$

(6.17)

where r_{max} is the maximum distance.

6.2.3 Theoretical Receiving Azimuth Angles and Maximum Receiving Powers

The positioning error and pointing error can be represented by the standard deviation of $\{\phi_l, r, \theta_T, \phi_T\}$. We can use the Monte-Carlo methods to obtain the root-mean-square error (RMSE) of the estimated ϕ_l, r, θ_T, and ϕ_T. The Monte-Carlo method first samples the receiving azimuth angles $\phi_{R,1}$ and $\phi_{R,1}$ to obtain ϕ_l, then samples the maximum receiving power $P_{R,1}$ and $P_{R,2}$ to obtain the distance r, θ_T, and ϕ_T. As we mentioned in Sect. 6.2.1, the receiving azimuth angle satisfies a Gaussian distribution with expectation $\phi_{R,i}$ and standard deviation $\sigma_{\phi_{R,i}}$. The standard deviation $\sigma_{\phi_{R,i}}$ is related to the employed measuring instrument. Without loss of generality, we set $\sigma_{\phi_{R,i}} = 1$. The theoretical receiving azimuth angle $\phi_{R,i}$ can be obtained by using the fact that $\boldsymbol{\mu}_{R,i}$ is perpendicular to $\boldsymbol{\mu}_\alpha$, i.e., $\boldsymbol{\mu}_\alpha \cdot \boldsymbol{\mu}_i = 0$, where $\boldsymbol{\mu}_\alpha$ can be obtained as

$$\boldsymbol{\mu}_\alpha = \boldsymbol{\mu}_T \times \boldsymbol{\mu}_l, \tag{6.18}$$

where $\boldsymbol{\mu}_T \triangleq [\cos\phi_T \cos\theta_T, \sin\phi_T \cos\theta_T, \sin\theta_T]^{\mathrm{T}}$ and $\boldsymbol{\mu}_l = [\cos\phi_l, \sin\phi_l, 0]^{\mathrm{T}}$.

Then using the relation $\boldsymbol{\mu}_\alpha \cdot \boldsymbol{\mu}_i = 0$, we can obtain the theoretical receiving azimuth angle $\phi_{R,i}$ as

$$\phi_{R,i} = \phi_l + (-1)^k \arcsin\left(\frac{\tan\theta_{R,i}}{\tan\theta_T} \sin\left(\phi_T - \phi_l\right)\right) + k\pi, \tag{6.19}$$

where $k \in \mathbb{Z}$. When $\phi_{R,i}$ is restricted in $[0, 2\pi]$, (6.19) has two possible results $\phi_{R,i}^1$ and $\phi_{R,i}^2$. To obtain the largest receiving power, we should choose the one with small gap $|\phi_{R,i} - \phi_l|$. Because $\left|\arcsin\left(\frac{\tan\theta_{R,i}}{\tan\theta_T} \sin\left(\phi_T - \phi_l\right)\right)\right| \leq \frac{\pi}{2} \leq \left|\pi - \arcsin\left(\frac{\tan\theta_{R,i}}{\tan\theta_T} \sin\left(\phi_T - \phi_l\right)\right)\right|$, we choose the theoretical $\phi_{R,i}$ as

$$\phi_{R,i} = \phi_l + \arcsin\left(\frac{\tan\theta_{R,i}}{\tan\theta_T} \sin\left(\phi_T - \phi_l\right)\right). \tag{6.20}$$

Then after obtaining the $\phi_{R,i}$, we can obtain $\theta_{r,i}$ by (6.15). The elevation angle θ_t of the transmitter in plane α can be obtained as

$$\theta_t = \arccos\left(-\cos\theta_T \cos\left(\phi_T - \phi_l\right)\right). \qquad (6.21)$$

Then the theoretical maximum receiving power $P_{R,i}$ for the ith receiver can be obtained by substituting r, θ_t and $\theta_{r,i}$ into (6.11). Then the expectation of the receiving photon numbers can be obtained as $\lambda_{s,i} = \frac{P_{R,i}T_d}{h\nu}$.

6.3 NLOS Ultraviolet Positioning Under Full-Duplex Communications

Now we extend the NLOS UV positioning method to the full-duplex communication scenario. A typical scenario of NLOS UV positioning in a full-duplex communication scenario is shown in Fig. 6.5, where user 1 is located at the origin of the coordinate system $C(x, y, z)$; l_T is the axis of the transmitted light beam of user 2; l_i is the axis of the receiving field of view (FOV) of user 1 intersecting with l_T under the elevation angle θ_i with the direction vector μ_i, where $i \in 1, 2$; l_I is the axis of the transmitted light beam of user 1 with the direction vector μ_I; l is the line between users 1 and 2 with the direction vector μ_l; ϕ_l is the angle between line l and the x-axis; r is the distance between users 1 and 2. When both users in the duplex communication system employ the same system design, it is reasonable to assume that user 1 knows the transmit power P_t of the transmitter of user 2, and the divergence angle β_t of the transmit beam. Due to the multiple scattering effects, there exists self-interference (SI) from the transmitted beam of user 1 on the ordinary channel between user 1 and user 2.

Fig. 6.5 Scenario of NLOS optical positioning in a full-duplex communication scenario

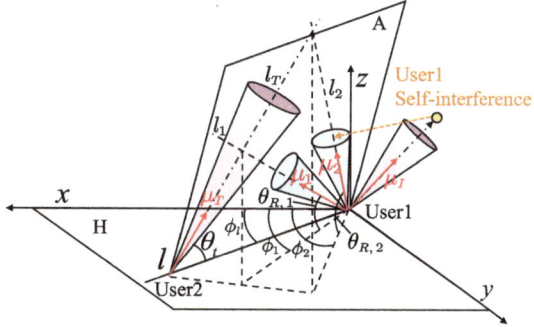

6.3.1 Approximation Model for SI

The presence of SI severely affects the signal detection at the receiver of user 1. Therefore, it is necessary to approximate the magnitude of SI by modeling the SI as an analytical form. In practical implementations, it is reasonable to assume that there is no first-order scattering signals in SI.

We use the Monte-Carlo integration model in Chap. 2 to simulate the amount of the SI [10]. To obtain the SI power, we rotate the receiver at different receiving azimuth angles with fixed elevation and azimuth angles of μ_I. We then use a Fourier series (FS) to fit the logarithm of SI power as

$$P_I(\phi) = 10^{a_0 + \sum_{n=1}^{N}[a_n \cos(\omega n\phi) + b_n \sin(\omega n\phi)]}, \tag{6.22}$$

where ϕ is the azimuth angle difference between the receiver and transmitter, a_0, a_n, and b_n are fitting parameters depending on the elevation angle difference θ, and N is the order of the fitted FS.

6.3.2 NLOS Optical Positioning Method Under SI

The received power at the receivers of user 1 consists of not only the signal power emitted by user 2 but also the SI power generated by the transmitter of itself. Therefore, $P_{R,i}$ consists of both the signal power $P_{S,i}$ and the SI power $P_{I,i}$ with $P_{R,i} \triangleq P_{S,i} + P_{I,i}$.

When user 1 rotates at an elevation angle θ_i, the received powers $P_{R,i}(\phi^1)$, $P_{R,i}(\phi^2)$, ..., $P_{R,i}(\phi^M)$ are measured at M azimuth angles $\phi^1, \phi^2, \ldots, \phi^M$. We define the difference between the receiving power $P_{R,i}(\phi^m)$ and SI power $P_I(\phi^m)$ as $P_{S,i}(\phi^m)$. The maximum signal power $P_{S,i}^{max}$ and its corresponding azimuth angle ϕ_i can be obtained by using the obtained $P_{R,i}(\phi^1)$, $P_{R,i}(\phi^2)$, ..., $P_{R,i}(\phi^M)$ and the fitted $P_{I,i}(\phi)$ as

$$\begin{cases} P_{S,i}^{max} = \max\{P_{R,i}(\phi^1) - P_{I,i}(\phi^1), P_{R,i}(\phi^2) - P_{I,i}(\phi^2), \\ \qquad\qquad \ldots, P_{R,i}(\phi^M) - P_{I,i}(\phi^M)\} \\ \phi_i = \underset{\{\phi^1, \ldots, \phi^M\}}{\arg\max}\{P_{S,i}(\phi^1), P_{S,i}(\phi^2), \ldots, P_{S,i}(\phi^M)\}. \end{cases} \tag{6.23}$$

According to (6.23), the estimated $\hat{\phi}_i$ is obtained as the corresponding azimuth angle of $\hat{P}_{S,i}$. Similar to Sect. 6.2, we assume that the receiving azimuth angle $\hat{\phi}_i$ satisfies a Gaussian distribution with mean ϕ_i and variance $\sigma_{\phi_{R,i}}^2$. After obtaining $P_{S,i}^{max}$ and ϕ_i, then we can obtain the distance r and azimuth angle ϕ_l similar to Sect. 6.2.

6.4 Numerical Results

In this section, we present some numerical results to verify the performance of the NLOS UV positioning using photon-counting receivers with and without SI. Unless otherwise specified, the parameters we adopt in this subsection are shown in Table 6.1.

Figure 6.6 presents the distance positioning error of distance under different distance r. The positioning error are obtained by using Monte-Carlo methods under two set of receiving elevation angles $\{10°, 80°\}$ and $\{20°, 70°\}$ with and without SI. From Fig. 6.6 we can see that the positioning error of distance will decrease when the gap between two receiving elevation angles increases. Besides, we can also observe that

Table 6.1 Simulation Parameters

Parameters	Value
$[r, \phi_l]$	$[50 \text{ m}, 0°]$
$[\theta_T, \phi_T]$	$[45°, 180°]$
$[\theta_I, \phi_I]$	$[45°, 0°]$
$[\theta_{R,1}, \theta_{R,2}]$	$[30°, 60°]$
$[\beta_t, \beta_r]$	$[10°, 30°]$
$[P_t, A_r]$	$[500 \text{ mW}, 1.77 \times 10^{-4}\text{m}^2]$
$[k_a, k_s^{Ray}, k_s^{Mie}]$	$[0.802, 0.266, 0.284] \text{ km}^{-1}$
$[\gamma, g, f]$	$[0.017, 0.72, 0.5]$
$[\sigma_{\phi_1}, \sigma_{\phi_2}]$	$[1°, 1°]$
T_d	1×10^{-5} s
h	6.626×10^{-34} J·s
c	2.998×10^{8} ms^{-1}
λ	260 nm

Fig. 6.6 Positioning performance under different distances

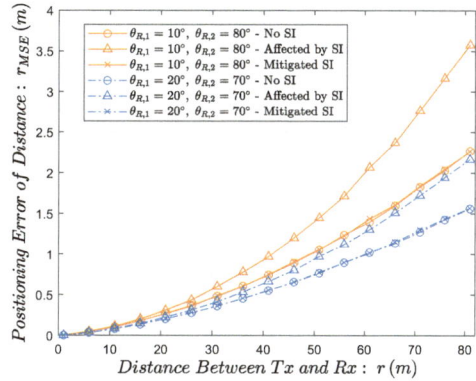

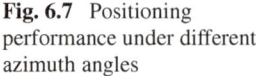
Fig. 6.7 Positioning performance under different azimuth angles

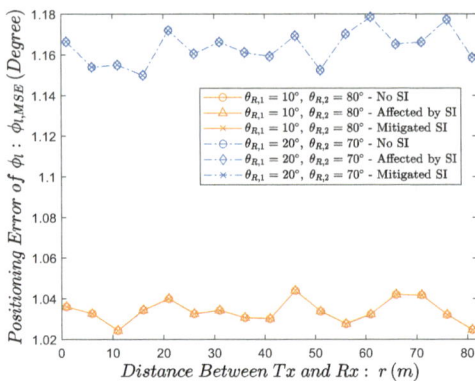

the SI can greatly degrade the positioning performance. However, the proposed positioning method under full-duplex UV communication can well mitigate the influence of SI.

Figure 6.7 presents the positioning error of azimuth angle under different distance r. From Fig. 6.7 we can see that the positioning error of azimuth angle will decrease when the gap between two receiving elevation angles increases. Besides, we can also observe that the SI can has little influence on the positioning performance of azimuth angle.

6.5 Summary and Future Directions

Traditional optical positioning methods are restricted by the line-of-sight scenarios. In this chapter, we have introduced a NLOS optical positioning method by using UV signals. Two NLOS UV positioning methods with and without SI based on the receiving powers are introduced and some numerical results are presented to verify the positioning performance of the NLOS UV positioning methods.

We have to remark that there still exists much room for improvement of the NLOS UV positioning. First, the number of receivers can be increased to avoid the possible multiple solution issues [12] to the optimization problem in Eq. (6.17). Second, the presented UV positioning methods are restricted in a two-dimensional positioning, which should be extended to a three-dimensional positioning in the future. At last, the introduced UV positioning methods in this chapter are based on the receiving power and ignored the time-of-arrival information. In the future, it is meaningful to explore some new NLOS UV positioning methods based on the channel impulse response, which can make use of both the receiving power and the time-of-arrival information.

References

1. Jovicic, A., Li, J., Richardson, T.: Visible light communication: opportunities, challenges and the path to market. IEEE Commun. Mag. **51**(12), 26–32 (2013)
2. Davidson, P., Piche, R.: A survey of selected indoor positioning methods for smartphones. IEEE Commun. Surv. Tutor. **19**(2), 1347–1370 (2016)
3. He, H., Ke, X., Zhao, T., et al.: Research of position in the wireless "solar-blind" ultraviolet mesh networks. Laser Technol. **34**(5), 607–610 (2010)
4. Zhao, T., Yu, Y., Bao, H., et al.: Ranging and positioning method using wireless solar blind ultraviolet. Opt. Precis. Eng. **25**(9), 2324–2332 (2017)
5. Li, Y., Wang, L., Xu, Z., et al.: Neighbor discovery for ultraviolet ad hoc networks. IEEE J. Sel. Areas Commun. **29**(10), 2002–2011 (2011)
6. Wang, L., Li, Y., Xu, Z., et al.: A novel neighbor discovery protocol for ultraviolet wireless networks. In: Proceedings of the 14th ACM International Conference on Modeling, Analysis and Simulation of Wireless and Mobile Systems, pp. 135–142 (2011)
7. Zhao, Y., Zuo, Y., Qin, H., et al.: A neighbor discovery protocol in ultraviolet wireless networks. In: 2014 Asia Communications and Photonics Conference (ACP), pp. 1–3 (2014)
8. Qi, H., Gong, C., Xu, Z.: Omnidirectional antenna array-based transmitter direction sensing in ultraviolet ad hoc scattering communication networks. In: 2019 IEEE International Conference on Communications Workshops (ICC Workshops), pp. 1–6 (2019)
9. Xu, Z., Ding, H., Sadler, B.M., et al.: Analytical performance study of solar blind non-line-of-sight ultraviolet short-range communication links. Opt. Lett. **33**(6), 1860–1862 (2008)
10. Yuan, R., Ma, J., Su, P., et al.: Monte-Carlo integration models for multiple scattering based optical wireless communication. IEEE Trans. Commun. **68**(1), 334–348 (2020)
11. Wang, S., Yuan, R., Peng, M., et al.: Non-line-of-sight ultraviolet positioning using two photon-counting receivers. In: Proceedings of the 2023 IEEE Global Communications Conference (GlobeCom2023), pp. 1–6 (2023)
12. Yuan, R., Wang, S., Liu, G., et al.: Non-line-of-sight ultraviolet positioning using linearly-arrayed photon-counting receivers. IEEE J. Sel. Areas Commun. Early Access (2024)

Chapter 7
Future Prospects of Ultraviolet Communications

Abstract The fast development of ultraviolet (UV) light sources and UV detectors has greatly promoted the research of non-line-of-sight (NLOS) UV communications in recent ten years. To achieve the practical implementation of NLOS UV communications, in this chapter, we present some future prospects of UV communications. In Sect. 7.1, we introduce the idea of integrated UV communication and positioning. Then we introduce the spatial diversity techniques of UV communications in Sect. 7.2. Then we introduce the main challenges of NLOS UV networking in Sect. 7.3. At last, we conclude this chapter in Sect. 7.4.

Keywords Integrated ultraviolet communication and positioning · Spatial diversity · Ultraviolet networking

7.1 Integrated Ultraviolet Communication and Positioning

7.1.1 Introduction to Integrated UV Communication and Positioning

The core factors currently restricting the communication speed and communication distance improvement of non-line-of-sight (NLOS) ultraviolet (UV) communication mainly include: (1) large pulse broadening of scattered signals, where different photons reach the receiver through different lengths of scattering paths, resulting in severe pulse broadening of UV signals; (2) high channel path loss, where photons are scattered once or multiple times before reaching the receiver, resulting in very weak received optical signals and limiting the improvement of communication distance. To address these issues, NLOS UV positioning technology can be introduced to locate the emitting light source and the transmitting geometry can be adjusted to maintain the single-scattering link and reduce the impulse spreading and path loss of the UV communication channel.

© The Author(s), under exclusive license to Springer Nature Singapore Pte Ltd. 2025 107
R. Yuan and Z. Wang, *Non-Line-of-Sight Ultraviolet Communications*,
SpringerBriefs in Computer Science, https://doi.org/10.1007/978-981-97-8543-8_7

Fig. 7.1 Typical scenario of integrated UV communication and positioning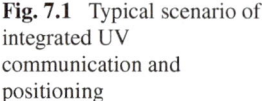

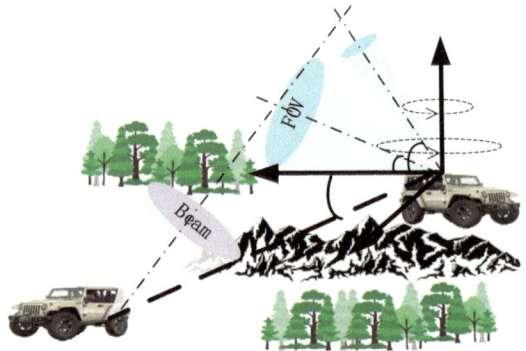

Thanks to the rapid development of UV LED technology and UV detection technology, it has become possible to achieve miniaturization and integration of UV communication terminals. However, current research on UV communication technology and UV positioning technology is relatively separate. In order to enhance the integration and collaboration of communication and positioning, it is necessary to study the integrated UV communication and positioning technology. Figure 7.1 shows a typical scenario of integrated UV communication and positioning.

7.1.2 Related Works in Integrated UV Communication and Positioning

Currently, the study of integrated communication and positioning mainly focuses on radio-frequency (RF) [1] or visible light wavebands [2]. The main contents include the aspects of the performance constrain between communication and positioning, the balance strategies between communication performance and positioning performance, and the performance boundaries of communication and positioning. These knowledge on RF and visible light wavebands can provide reference for the research of integrated UV communication and positioning technology. For example, in the visible light band, Yang et al. proposed an integrated visible light communication positioning system based on low complexity suboptimal subcarriers, and used sequence quadratic programming method to solve the nonlinear power allocation problem, which can effectively improve the system's speed and positioning accuracy [3]. Chen et al. proposed an integrated framework for visible light communication positioning based on CDMA, which can achieve a bit-error rate of 1.8×10^{-3} and a positioning accuracy of 1.50 centimeters in a real-time system with 4 users [4]. Jin et al. proposed a new demodulation scheme for a mobile visible light communication positioning integrated system based on adaptive feedback threshold, which can achieve high reliability communication and high-precision positioning of mobile integrated very low frequency communication systems [5]. Celik utilizes visible light positioning technology to obtain positioning data that improves communication link quality and increases received signal-to-noise ratio [6]. Cao et al. proposed an integrated waveform for communication ranging to balance laser ranging performance and high-speed communication requirements [7]. However, the above-mentioned studies [3–7] mainly focused on the allocation of communication and positioning resources, and failed to analyze the constraint relationship between the performance limits of communication and positioning.

The analysis on the constraint relationship between the performance limits of communication and positioning has been extensively studied in the RF band. For example, Guo et al. derived the relationship between communication mutual information and perceived minimum mean-square error in Gaussian channels [8]. Sutivong et al. studied the relationship between communication capacity and channel state estimation performance in Gaussian channels when channel state is correlated, and provided the optimal equilibrium feature between information rate and channel state perception mean square error [9]. Liu et al. analyzed the optimal beamforming design problem in the integrated scenario of radar perception and communication based on the Cramer-Rao boundary [10]. However, the above studies [8–10] are based on Gaussian channels, which have significant differences in weak link characteristics with ultraviolet light, making it difficult to directly apply RF band related research to scattered light channels.

7.1.3 Challenges in Integrated UV Communication and Positioning

The current research on the integrated UV communication and positioning mainly faces the following challenges:

- The evaluation of the performance limit of UV communication and positioning under scattered channels is difficult.
- The constraint relationship between UV communication and UV positioning performance is unclear.
- The node design for integrated UV communication and positioning should consider the self-interference effect between the communication and positioning links.

7.2 Spatial Diversity Techniques of UV Communications

7.2.1 Introduction to Spatial Diversity Techniques of UV Communications

Space diversity technology is a commonly used method for combating turbulent channel fading in free-space optical communication systems. For short-range UV communication systems, the influence of turbulence is often negligible, and the fluctuation of the received signal mainly comes from the Poisson variance of the optical signal caused by high channel loss. The intensity fluctuation of this weak light signal can be regarded as a special fading effect, so spatial diversity technology can also be used to improve the channel capacity of the short-rang UV communication system and increase the signal reception gain [11]. In addition, the emission power of existing single UV LEDs and the reception area of single ultraviolet detectors are limited. To obtain greater emission power and reception signal strength, a multiple-input multiple-output communication structure is often required. In this context, the combining method of multiple output signals at the receiver end will have a significant impact on the subsequent detection performance.

The maximum-ratio combining (MRC) and equal-gain combining (EGC) methods are widely adopted combining techniques in traditional optical communication with spatial diversity. The MRC method is the optimal combining method in terms of maximizing the receiving signal-to-noise ratio (SNR). However, for UV communication, the fluctuation intensity of weak light signals increases with the increase of signal strength. In this context, the MRC method may no longer be the optimal merging method for maximizing the SNR. A typical single-input multiple-output NLOS UV communication configuration is shown in Fig. 7.2. Multiple photodetectors (PDs) are employed to detect the UV signal from different sub-channels. The time-of-arrivals at different PDs are different from each other due to different receiving geometries. Besides, due to the weak link property, the received signals are determined by a Poisson distribution. Therefore, it is necessary to explore the optimal combining method for spatial diversity technology of NLOS UV communication under weak links.

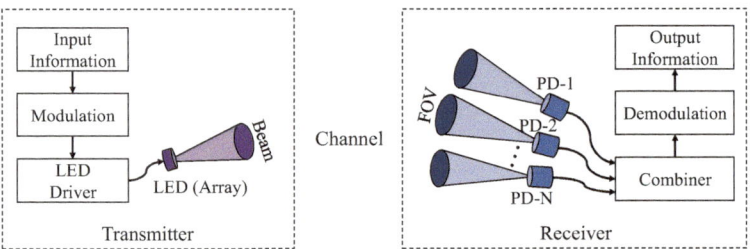

Fig. 7.2 Typical SIMO UV communication system

7.2.2 Related Works in Spatial Diversity Techniques of UV Communications

. Currently, there are some pioneering theoretical and experimental works on spatial diversity receptions of UV communication systems [12–14]. For example, Han et al. proposed the first diversity reception technique for NLOS UV communication in [12], where a dual-PMT diversity receiver using EGC method was employed. Guo et al. employed a combining method with combining weight equalling the channel gain [13]. Meng et al. experimentally studied a 1×4 real-time single-input multiple-output (SIMO) scheme with EGC for NLOS UV communication system [14]. However, these works [12–14] adopted the additive white Gaussian noise (AWGN) assumption for the detectors, which may fail for NLOS UV communication systems under weak links.

There are also some works considered the Poisson detecting properties instead of AWGN assumption in the diversity reception of NLOS UV systems [15–18]. For example, El-Shimy et al. developed a spatiotemporal channel model for multi-element imaging receivers of NLOS UV communications in [15], where the EGC as well as the subset combining was considered. Qin et al. proposed an analytical link bandwidth model considering the Poisson detecting properties for an SIMO UV communication system in [16]. In [17], a linear minimum mean-square error SIMO receiver was proposed for UV communication systems, where the linear conversion method was generalized to a non-linear conversion in [18]. However, theoretical analyses in [15–18] have inherently assumed a photon-counting mode for receivers, while photon-countering receivers are challenging to achieve in practical implementations. Besides, the turbulent fading effect was ignored in above works.

7.2.3 Challenges in Spatial Diversity Techniques of UV Communications

The current research on the spatial diversity techniques of UV communications mainly faces the following challenges:

- The optimal combining method of the spatial diversity for the NLOS UV communication is not clear.
- The analysis of the impact of the non-AWGN noise on the combining method for NLOS UV communication is challenging.
- The impact of the turbulence on spatial diversity techniques should be analyzed for NLOS UV communication under large communication ranges or elevation angles.

7.3   Non-Line-of-Sight Ultraviolet Networking

7.3.1   Introduction to Non-Line-of-Sight Ultraviolet Networking

Currently, the research on UV communication mainly focused on the point-to-point communication scenarios. However, practical implementations of UV communication required the development of UV networking. Due to the weak link property of UV communication, short-rang UV communication is still a major communication scenario. Besides, the major application of UV communication also requires the development of mobile UV communication. In this context, the ad-hoc communication network becomes the most promising networking technique for UV communications.

The key research points of UV ad-hoc network include the inter-link-interference issue of the physical layer, the media access control (MAC) protocol of the data-link layer, and the routing protocol of the network layer. First, in the physical layer, UV signals are scattered by atmospheric particles and aerosols before reaching the receiving end, forming a NLOS UV link. Therefore, different NLOS UV links in an UV ad-hoc networks will interfere each other and cause the inter-link-interference issue. Second, in the data-link layer, due to the geometrical constrains of the transmitter and receiver, the transmitting and receiving of UV signals are usually not omnidirectional, resulting in a "deafness effect" in the UV ad-hoc network. As shown in Fig. 7.3, there are three typical transmitting and receiving configurations of UV NLOS communication, i.e., the transmitting-omnidirectional-receiving-omnidirectional (TORO) configuration, the transmitting-directional-receiving-omnidirectional (TDRO) configuration, the transmitting-omnidirectional-receiving-directional (TORD) configuration, and the transmitting-directional-receiving-directional (TDRD) configuration.

Fig. 7.3 Typical transmitting and receiving configurations of UV communication

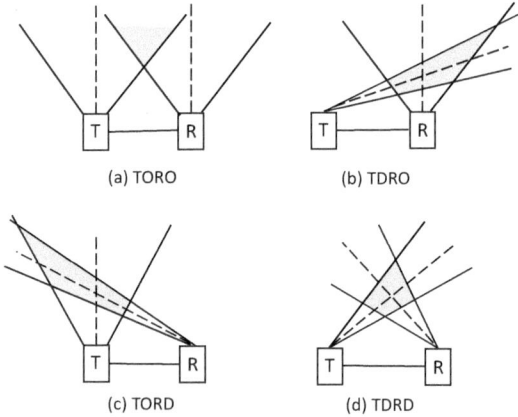

From Fig. 7.3 we can see that the "deafness effect" is inevitable as long as the receiving end is not omnidirectional. Therefore, it is necessary to study special MAC protocol for UV ad-hoc network considering the deafness effect. At last, in the network layer, due to the mobile property of UV nodes and the flexibility of the transmitting and receiving geometries, it is challenging to study the routing protocol of UV ad-hoc network. In a word, the characteristics of the UV scattering signals require a new design on either the physical layer, data-link layer, or network layer of the UV ad-hoc network.

7.3.2 Related Works to Non-Line-of-Sight Ultraviolet Networking

In the physical layer, some preliminary work has been performed to study the inter-link-interference issue of UV ad-hoc network. For example, Zhao et al. investigated the influence of geometrical parameters such as the receiving elevation angle and field-of-view angle at the receiving end on the inter-link-interference of UV network [19]. Jiang et al. studied the influence of communication parameters such transmitting power and transmitting bit rate on the inter-link-interference of UV network [20]. In avoid the inter-link-interference, Li et al. proposed the time-division multiplexing technology by allocating different time slots to the possible interfering nodes and derived an expression of the power saving rate [21]. Song et al. analyzed the impacts on different transmitting and receiving parameters on the performance of UV network and concluded that the inter-link-interference can be reduced by choosing appropriate geometrical parameters [22]. Besides the interference issue, the connectivity issue is another important issue in UV network. Vavoulas et al. studied the impacts of parameters such as node density, transmission power, and data rate on the connectivity of UV ad-hoc networks [23]. Zhao et al. found that the connectivity performance of pulse-position modulation (PPM) based UV network is higher than that of on-off keying modulation (OOK) based UV network [24]. Li et al. found that the connectivity of TDRD configuration is better than that of TORO or TDRO configuration [25]. Qi et al. also found that directional antenna arrays are superior to omnidirectional antennas in connectivity [26].

In the data-link layer, based on the TDRO configuration, Ke et al. proposed an angle aware directional MAC protocol, which utilizes the directionality of TDRO configuration and dynamically selects the optimal transmitting elevation angle through an angle aware mechanism, thereby reducing transmission hops and avoiding the "deafness effect" [27]. Zhang et al. studied the "deafness effect" problem in TDRD configuration and proposed UV communication directional MAC protocol based on spatial division multiplexing [28]. Ke et al. proposed a new backoff algorithm based on backoff counter values to address the issue of unfair channel resource occupation between nodes in UV ad-hoc networks [29]. Li et al. fully considered the physical layer characteristics of UV communication and proposed a UVOC-MAC protocol, which utilizes the NLOS links for spatial reuse, adaptively changing data rates based on channel delay characteristics, and selects full-duplex or half-duplex modes based on pointing angles [30]. Li et al. proposed a lossless contention MAC protocol based on the characteristics of the UV physical layer (UVLLC-MAC), which utilizes the optical power superposition mechanism of the UV physical layer to assign different priorities to different data services [31].

In the network layer, Song et al. proposed a routing protocol for UV space division multiplexing ad-hoc networks based on the multicast ad-hoc on-demand distance vector (MAODV) protocol, and the effectiveness of the routing protocol was verified through simulation [32]. Chen et al. proposed an ad-hoc on-demand multipath distance vector (AOMDV) routing algorithm for UV ad-hoc networks to improve the performance of the MAODV protocol [33]. Qiu et al. proposed an improved ant colony algorithm, i.e., the UV ant colony optimization (UVACO) algorithm, for routing protocols to address issues such as high network transmission latency and uneven node energy consumption in UV network [34]. Zhang et al. proposed a weighted routing protocol based on link state awareness to address the problem that traditional DSR routing protocols cannot adapt to rapid changes in network topology in high-speed mobile airborne UV communication network scenarios [35]. Zhao et al. proposed a wireless UV cooperative drone swarm energy efficiency optimization routing algorithm to address the impact of strong interference and complex environments on drone communication [36].

7.3.3 Challenges in Non-Line-of-Sight Ultraviolet Networking

The current research on the NLOS UV networking mainly faces the following challenges:

- The impact of the inter-link-interference issue on the networking performance, e.g., the throughput is still not well studied.
- The impact of the "deafness effect" due to the system geometry should be further explored.
- The impact of the scattering effect of the UV signals on the routing strategy has not been studied.

7.4 Summary

In this chapter, we introduced three future directions of UV communications, including the integrated UV communication and positioning, the diversity techniques, and the UV networking. The basic ideas of these research directions are explained and the related works are briefly introduced. We also summarized some major challenges met in these research directions.

References

1. Jovicic, A., Li, J., Richardson, T.: Visible light communication: opportunities, challenges and the path to market. IEEE Commun. Mag. **51**(12), 26–32 (2013)
2. Liu, F., Cui, Y., Li, J., Masouros, C., et al.: Integrated sensing and communications: toward dual-functional wireless networks for 6G and beyond. IEEE J. Sel. Areas Commun. **40**(6), 1728–1767 (2022)
3. Yang, H., Chen, C., Zhong, W., et al.: Resource allocation for multi-user integrated visible light communication and positioning systems. In: Proceedings of 2019 IEEE International Conference on Communications (ICC), pp. 1–6. IEEE (2019)
4. Chen, D., Fan, K., Wang, J., et al.: Integrated visible light communication and positioning CDMA system employing modified ZCZ and Walsh code. Opt. Express **30**(22), 40455–40469 (2022)
5. Jin, J., Lu, H., Wang, J., et al.: Adaptive feedback threshold-based demodulation for mobile visible light communication and positioning integrated system. Opt. Express **30**(8), 13331–13344 (2022)
6. Celik, Y.: Indoor visible light communication and positioning with pan-tilt receiver. In: Proceedings of 2022 Innovations in Intelligent Systems and Applications Conference (ASYU), pp. 1–5 (2022)
7. Cao, M., Wang, Y., Zhang, Y., et al.: A unified waveform for optical wireless integrated sensing and communication. In: 2022 Asia Communications and Photonics Conference (ACPC), pp. 448–452 (2022)
8. Guo, D., Shamai (Shitz), S., Verdu, S.: Mutual information and minimum mean-square error in Gaussian channels. IEEE Trans. Inf. Theory **51**(4), 1261–1282 (2005)
9. Sutivong, A., Chiang, M., Cover, T.M., et al.: Channel capacity and state estimation for state-dependent gaussian channels. IEEE Trans. Inf. Theory **51**(4), 1486–1495 (2005)
10. Liu, F., Liu, Y.F., Li, A., et al.: Cramer-Rao bound optimization for joint radar-communication beamforming. IEEE Trans. Signal Process. **70**, 240–253 (2022)
11. Yuan, R., Peng, M.: Single-input multiple-output scattering based optical communications using statical combining in turbulent channels. IEEE Trans. Wirel. Commun. **23**(4), 2560–2574 (2023)
12. Han, D., Liu, Y., Zhang, K., et al.: Theoretical and experimental research on diversity reception technology in NLOS UV communication system. Opt. Express **20**(14), 15833–15842 (2012)
13. Guo, L., Meng, D., Liu, K., et al.: Experimental research on the MRC diversity reception algorithm for UV communication. Appl. Opt. **54**(16), 5050–5056 (2014)
14. Meng, X., Zhang, M., Han, D., et al.: Experimental study on 1×4 real-time SIMO diversity reception scheme for a ultraviolet communication system. In: proceedings of 2015 20th European Conference on Networks and Optical Communications (NOC), pp. 1–4. IEEE (2015)
15. El-Shimy, M.A., Hranilovic, S.: Spatial-diversity imaging receivers for non-line-of-sight solar-blind UV communications. J. Light. Technol. **33**(11), 2246–2255 (2015)
16. Qin, H., Zuo, Y., Li, F., et al.: Analytical link bandwidth model based square array reception for non-line-of-sight ultraviolet communication. Opt. Express **25**(19), 22693–22703 (2017)
17. Gong, C., Xu, Z.: LMMSE SIMO receiver for short-range non-line-of-sight scattering communication. IEEE Trans. Wirel. Commun. **14**(10), 5338–5349 (2015)
18. Shen, Z., Ma, J., Su, P.: LMMSE-based SIMO receiver for ultraviolet scattering communication with nonlinear conversion. IEEE Wirel. Commun. Lett. **10**(10), 2140–2144 (2021)
19. Zhao, T., Zhang, A., Jin, D., et al.: Research on the inter-link interference model in wireless ultraviolet non-line-of-sight communication. Acta Optica Sinica **33**(7), 0706023 (2013)
20. Jiang, X., Luo, P., Zhang, M.: Performance analysis of non-line-of-sight ultraviolet communications with multi-user interference. In: Proceedings of 2013 IEEE/CIC International Conference on Communications in China (ICCC), pp. 199–203. IEEE (2013)
21. Li, J., Wu, X., Wang, H., et al.: Anti-interference relayed link method and power requirement analysis for ultraviolet non-line-of-sight communication. Laser Optoelectron. Prog. **52**(3), 030601 (2015)

22. Song, P., Ke, X., Song, F., et al.: Multi-user interference in a non-line-of-sight ultraviolet communication network. IET Commun. **10**(13), 1640–1645 (2016)
23. Vavoulas, A., Sandalidis, H., Varoutas, D.: Connectivity issues for ultraviolet UV-C networks. J. Opt. Commun. Netw. **3**(3), 199–205 (2011)
24. Zhao, T., Wang, Y., Gao, Y.: Research on connectivity performance of wireless UV non-line-of-sight communication networks. Optoelectron. Laser **2015**(1), 68–74 (2015)
25. Li, C., Li, J., Xu, Z., et al.: Study on the k-connectivity of UV communication network under the node distribution of RWP mobility model in the arbitrary polygon area. IEEE Photon. J. **12**(4), 1–12 (2020)
26. Qi, H., Zou, D., Gong, C., et al.: Two-dimensional intensity distribution and connectivity in ultraviolet ad-hoc network. In: proceedings of 2020 IEEE International Conference on Communications (ICC), pp. 1–6. IEEE (2020)
27. Ke, X., Chen, J., Hou, Z.: Directional media access control protocol for UV non-line-of-sight communication. J. Optoelectron. · Laser **2011**(8), 1190–1195 (2011)
28. Zhang, X., Zhao, S., Li, Y., et al.: Directional MAC protocol for multi-channel UV-optical communication based on space division multiplexing. Laser Technol. **40**(3), 451–455 (2016)
29. Ke, X., He, H., Cheng, X.: A new backoff algorithm of MAC layer in UV ad-hoc communication network. J. Optoelectron. · Laser **21**(7), 1002–1006 (2010)
30. Li, Y., Ning, J., Xu, Z., et al.: UVOC-MAC: A MAC protocol for outdoor ultraviolet networks. In: proceedings of the 18th IEEE International Conference on Network Protocols, pp. 72–81. IEEE (2010)
31. Li, C., Xu, Z., Li, J., et al.: Performance of the UV multi node network under the lossless contention MAC protocol. IEEE Photon. J. **14**(2), 1–7 (2022)
32. Song, X., Song, F., Song, P., et al.: Routing protocol of ultraviolet space division multiplexing ad-hoc network. Chin. J. Lasers **44**(10), 1006005 (2017)
33. Chen, C., Song, P., Li, Y., et al.: On-demand multipath distance vector routing algorithm for UV-AD hoc networks. J. Xi'an Polytech. Univ. **33**(1), 74–80 (2019)
34. Qiu, D., Li, J., Wang, J., et al.: An improved ant colony algorithm for ultraviolet communication networks. Opt. Commun. Technol. **48**(2), 12–17 (2019)
35. Zhang, X., Wang, Y.: A weighted routing protocol for airborne ultraviolet communication network based on link state perception. In: proceedings of the International Conference on Advances in Computer Technology, Information Science and Communication (CTISC), pp. 103–106. IEEE (2021)
36. Zhao, T., Cheng, M., Zhang, G., et al.: Energy efficiency optimization routing algorithm for wireless ultraviolet cooperative UAV swarm. Laser Optoelectron. Prog. **59**(5), 0506005 (2022)

Index